SOTHIS PRESS

Sothis Press aims to completely reimagine the public domain book with an emphasis on universal access to captivatingly designed print editions. We believe that technology will enable every single book ever published to be perpetually accessible in beautiful print form. Sothis is beauty in ubiquity.

www.sothis.press
1 Waterhouse Square, London, ECIN 2ST

Building Plans and How to Draw Them

A Simple Series of Practical Lessons on Architectural Drawing, Showing Every Step Necessary to Draw the Full Working Plans of Buildings; Intended for the Self-Instruction of Building Mechanics

Isaac P. Hicks

SOTHIS PRESS

ISBN 978-1-0365-0572-1

Building Plans and How to Draw Them
is in the public domain.

This edition is published by
Sothis Press © 2023. The company
is registered in England and Wales.
The Company Number is 14900389.

BUILDING PLANS

AND HOW TO DRAW THEM

A SIMPLE SERIES OF PRACTICAL LESSONS ON ARCHITECTURAL DRAWING, SHOWING EVERY STEP NECESSARY TO DRAW THE FULL WORKING PLANS OF BUILDINGS

INTENDED FOR THE SELF-INSTRUCTION OF BUILDING MECHANICS

By I. P. HICKS

Author of "Architectural Perspective for Beginners," "Builders' Guide," "Estimators' Price Book," etc., etc.

FULLY ILLUSTRATED

THIRD EDITION

NEW YORK
INDUSTRIAL PUBLICATION COMPANY
1909

BUILDING PLANS AND HOW TO DRAW THEM

CONTENTS

PREFACE

The pages which comprise this work have been prepared with a great deal of care, in the belief that there is need of a thorough treatise on architectural drawing for mechanics, showing the learner how to proceed step by step in every detail of the work.

The examples given as lessons in drawing embrace a wide range of work, and it has been the aim of the author to select only such figures for practice as are most likely to meet the wants of working mechanics.

The popularity with which Hicks' Builders' Guide has been received has brought many letters of inquiry which have been convincing proof of the wants and needs of mechanics for a work of this kind. No pains have been spared to make it practical for those who desire to profit by self instruction. Every subject treated has been made plain and easy to comprehend, and it is believed the work will prove the keynote to success for those desiring instruction in this branch of the building trade.

To one seeking information of a practical nature, the selection and arrangement of the matter within these pages cannot fail to supply the student with a fund of knowledge not easily obtained in any other work. Thanking the readers for their many kind and favorable expressions of the work as published serially in the columns of *Carpentry and Building*, it is now condensed into convenient space for guidance and instruction, and is committed to their use, trusting that it will be favorably received.

THE AUTHOR.

OMAHA, NEB., 1897.

DRAWING INSTRUMENTS

The subject of architectural drawing is one of no small interest to the practical mechanic in whatever branch of the building trade he may be engaged. Carpenters, contractors and builders are often called upon to furnish plans, especially of medium and low cost buildings which are not considered of sufficient importance to warrant employing the services of a professional architect. It is in connection with this class of work that a practical knowledge of drawing is particularly valuable to the building mechanic, for he who understands drawing sufficiently to take a rough sketch of a floor plan and draw it up in practical shape, presenting the corresponding elevation with the details necessary to carry out the design in a comprehensive manner, is the man who, as a rule, readily secures and retains the public confidence. The one who understands and can make the working drawings for buildings has a decided advantage over a competitor possessing little or no skill as a draftsman. The ability to draw up a plan after the manner of an architect has the effect to gain for the draftsman the favor and confidence of his patrons in a way that seems to assure them he understands his business, and therefore is the more competent to do the work. Thus in cases where competency is required the lowest bidder does not always stand the best chance in the race for a job. This, and the fact that a knowledge of drawing is necessary to enable workmen to take a set of architects' plans and work from them understandingly, has brought about a desire on the part of many to learn the art of drawing.

A brief and comprehensive treatise on the subject, giving the practical instructions necessary to carry the work to a successful issue, has long been felt to be desirable. Many books and articles have been written on the subject, but for the most part they have been far from meeting the wants of carpenters and contractors engaged in the general building trade. Many have been disappointed in books, partly because they have not been as practical as could be wished, but more especially because the readers failed to put into practice the principles and ideas which the books illustrated and described.

In order to obtain the full benefit from a work of instruction on any subject, it is necessary to verify what is taught by practice, and this is assuredly the fact in regard to drawing. Practice is the most essential point toward the advancement and success of the draftsman in the art of drawing, and without it the best books and articles are likely to prove a failure. It will not answer to merely read the instructions and to look at the examples given; a few ideas may be gained by this course, but to acquire the art and become proficient in it will require practice. It is the only successful method to pursue, and, therefore, I say to those who are seeking instructions in drawing, take up the rule, pen and pencil and begin practice at once by making the drawings given in connection herewith as lessons in drawing.

It is not necessary to illustrate and describe all the tools and appliances used in drafting, as this has been done time and again, until every carpenter and contractor is familiar with them; therefore, I will only describe a few of the most essential tools and appliances used in practice. A drawing board, 16 x 24 inches, will be large enough for convenience in making the drawings given as lessons, and also large enough for many other purposes. For be it known that a large drawing board for small work is more or less inconvenient and undesirable. In my opinion it is better to have two or three sizes to meet the varying requirements of large and small work.

Drawing boards should be made of soft wood, with all angles right angles, so that the T square will work properly from any one of its four sides. The T square is an indispensable tool in the draftsman's outfit, and it is very desirable to have several sizes. A 20-inch blade will meet the requirements of these lessons, but in practice much longer blades will often be required. The cost of a good T square, 20-inch blade, is not great—say about $1—and in selecting one I would suggest Deane's patent adjustable head. Triangles are very handy and convenient to use in connection with the T square and can be employed to advantage in many ways. Triangles may be had for 25 to 75 cents each, according to size.

In the way of other tools I would recommend a small case of German silver instruments of a good quality, containing 4½-inch dividers, with pen and pencil points and lengthening bar, drawing pen and protractor, the cost of which does not need to exceed $4, although the draftsman can, if he desires, procure a $15 set of instruments.

Drawing rules are among the necessities that go to make up the draftsman's outfit, and of these he should have several at his disposal. The architect's triangular scale is the king of all rules for setting off the scale measurements of drawings, after which they may be finished by drawing the lines with the aid of the T square, triangles, common rules or straight edges. I do not advocate a general use of the triangular scale with which to draw lines, for the reason that it is a costly rule, very finely and accurately marked, and to make a general use of it in drawing lines would have a tendency to obscure the fine divisions of the rule and thus impair its usefulness in setting off accurate measurements. The rule should never be used in drawing ink lines, but can be used for drawing the pencil lines necessarily required in outlining a plan, if it is desirable to do so. These remarks are more especially to warn the draftsman against the indiscreet use of a costly rule, which might soon impair its usefulness or spoil it entirely. The cost of triangular scales is from $1.50 to $5. They are made in 12-inch and 24-inch lengths, the 12-inch length being the one in general use. Thus it will be seen that the draftsman's outfit for drawing need not cost more than $7 to $10. The drawing boards, T squares, triangles and straight edges he can make for himself, thus confining the expense to the case of instruments, the triangular scale and common rules. The author began practicing with a 15-cent outfit, consisting of a common rule and lead pencil. By the aid of triangles which I made myself, I was enabled to square drawings from the rule, draw the horizontal and plumb lines and finish up the drawing comparatively easy. I merely mention this so beginners will not get the idea that a costly outfit of tools is required; but to all who can afford it I would say do not retard your progress in drawing by trying to get along without the necessary tools, as it is a waste of time and poor economy.

In reference to the triangular scale, it is necessary that the draftsman should become familiar with the different scales represented. This necessity will not be felt so much in copying the drawings here given as lessons as in the making of original drawings. In this the draftsman will soon discover that mental calculations are constantly coming up, and to compute the distances and set them off readily and accurately on the drawing requires a thorough knowledge of the drafting scales. Before proceeding with the instructions in drawing we will give a brief description of the triangular scale and its uses. By referring to Fig. 1 the shape of the triangular scale will be plainly seen. In shape an end section presents the form of an equilateral triangle, traversed lengthwise with a semicircular groove. This particular shape gives the rule six planes or surfaces, on which the different scales are marked. The shape of the rule permits of the marking being plainly visible to the draftsman, and is specially adapted for making the scale measurements of drawings with the greatest accuracy. On the six planes of the rule are marked 11 scales, which we will take in their regular order. First one plane is graduated to sixteenths, which, of course, can be used in drafting to represent the scale of 1-16 inch to the foot. The other ten scales are marked on the remaining five planes in pairs and in the following order, 3-32, 3-16, ⅛, ¼, ⅜, ¾, ½, 1 inch, 1½ inches and 3 inches to the foot.

Placing the rule in position for drawing or setting off measurements, we very naturally use the opposite edge from the one facing us. This will be plainly seen by taking a rule and placing it in the position of Fig. 1, which will bring the scales of 3-32 and 3-16 in proper position by which to draw lines and set off measurements on the drawing. When in this position the scale of 3-32 inch to the foot will be found figured on the plane from left to right and the scale of 3-16 inch to the foot, which is a multiple of 3-32, will be found figured from right to left, in the semicircular groove just above the plane. The scales each have at their starting point 3-32 and 3-16 inch respectively, graduated to sixths and twelfths, for setting off fractional parts of a foot, as in representing inches.

Fig. 1.—Broken View of a Triangular Scale.

Referring to Fig. 1, we find the following scales shown: At the left end, top plane, 1½ inches to the foot; right end, same plane, 3 inches to the foot; left end, bottom plane, ½ inch to the foot and right end, same plane, 1 inch to the foot. Thus it will be plainly seen that the larger scales are just twice the smaller, and being represented on the same planes the graduations of the smaller scale represent halves of the larger scale and thus work in harmony when properly understood and applied. It will be noticed that the figures of the ½-inch and 1-inch scales appear wrong side up, but when the rule is turned end for end, as it should be used in drawing by these scales, the figures will appear right and as follows: The ½-inch scale will be found figured from right to left on the plane and the 1-inch scale figured from left to right just above the plane in the semicircular groove. Thus in setting off distances by the inch scale, if ½ foot was wanted the draftsman could readily locate it by the divisions of the ½-inch scale figured on the plane. All the other scales on the rule are figured in like manner. The scales being figured differently and in opposite directions the liability of making mistakes or getting the scales mixed is avoided.

In setting off measurements from any of the scales always start from the figure 0 of the desired scale for even feet, and if a fractional part of a foot is wanted, count off such part on the fine divisions of the scale and start from this point, then the figures on the scale will represent the feet and inches. For example, the ½-inch scale has ½ inch graduated as follows: 1. The long lines divide the scale into fourths, which represent 3 inches. 2. The next longest lines divide the scale into twelfths, which represent inches. 3. The short lines divide the scale into twenty-fourths, which represent ½ inches. Thus by this scale the draftsman can carry his scale measurements to a point indicating ½ inches. Now if it were required to set off 6 feet 9 inches by the scale, count off 9 inches from the 0 mark on the scale, which will be the third long line to the right, then start from this point with the measurement and follow the scale marked on the plane to the left to 6, which will give the measurement of 6 feet 9 inches. Proceed in like manner in setting off measurements by any of the scales found on the rule. It is not necessary to describe the graduations of all the scales, as to understand the working of one is sufficient to enable the draftsman to readily become familiar with the divisions of them all, as they all embody the same principle, and it is just as easy to draw by one scale as the other, with the exception of cases where the fine scales bring lines so close together that they cannot be drawn distinctly.

We will only describe the graduations of the 3-inch and ¼-inch scales to show the contrast and what it is possible to accomplish with the scales. The 3-inch scale will be found graduated as follows: 1. The long lines divide the scale into twelfths, which represent inches.

2. The next longest lines divide the scale into twenty-fourths, which represent half inches. 3. The next longest lines divide the scale into forty-eighths, which represent quarter inches. 4. The short lines divide the scale into ninety-sixths, which represent eighth inches. Thus by the scale of 3 inches to the foot the draftsman can carry his scale measurements to a point indicating the eighth part of an inch, which is as fine as required in ordinary practice.

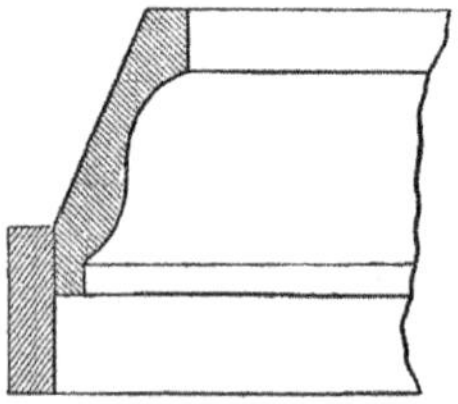

Fig. 2.—Crown Molding and Fascia Drawn to a Scale of 3 Inches to the Foot.

Referring now to the ¼-inch scale, which is the one generally used in making elevation drawings, we find it graduated as follows: The long lines divide the scale into halves, which represent 6 inches; the next longest lines divide the scale into quarters, which represent 3 inches; the short lines and finest divisions of the scale divide it into twelfths and represent inches. Thus by this scale the draftsman can carry his scale measurements to a point indicating the inches in elevation drawings, which is about as fine as it is possible to draw the lines. The architects' triangular scale is calculated for fine work, and with it the finest calculations in drawing can be made. As previously shown, the fine divisions of the ¼-inch scale graduated to twelfths make it possible in drawing by the scale to represent an inch on the plan. By the division of ¼ inch into 12 parts an inch would represent 48 parts; hence one of these parts is really the forty-eighth part of an inch, and comes very nearly, if not quite, being finer than the ordinary draftsman can distinctly draw the lines. Thus it will be seen that in drawing elevations great care will be necessary to keep all parts in accordance with the scale. For example, to represent the fascia and crown molding of a cornice in its usually limited space requires some fine work. The space usually required for the fascia and crown

Fig. 3.—Crown Molding and Fascia Drawn to a Scale of ¼ Inch to the Foot.

molding of a cornice on the average dwelling is from 5 to 7 inches. In this case we will call it 6 inches to make it easy for beginners. According to the scale, 6 inches would be represented by ⅛ inch in the drawing. Now we have only ⅛ inch space in which to draw the lines necessary to represent the fascia and crown molding, and as each member of the fascia and molding requires a line to properly show its profile, it would require at least five parallel lines to represent a very plain fascia and crown molding. To draw all these lines in the allotted space is about as fine work as the average draftsman is capable of doing. For example, we will draw the fascia and molding on a scale of 3 inches to the foot in order to distinctly show the different members. Referring to Fig. 2, it will be seen that it requires 1½ inches space to show the profile of fascia and molding. To show all these lines in an elevation in the small space of ⅛ inch is no small task, and this is only one example out of many that arise in architectural drawing.

Fig. 3 represents the fascia and molding drawn to ¼-inch scale. In drawing elevations occasionally a line is left out and slight deviations are sometimes made. This is the reason details drawn to larger scales are necessary, in order that the workmen may be enabled to carry out the design to the true meaning and intent of the architect. In making the elevations to the ¼-inch scale the draftsman should work to the scale and keep the proper proportions as much as possible, and such parts as cannot be properly shown in the elevations should be represented in the details drawn to larger scales.

A FIVE-ROOM COTTAGE.

In the selection of examples to serve as lessons in drawing we have chosen for the subject the plan of a five-room cottage of medium cost. This selection has been made in view of the fact that the architectural drawing of medium and low cost dwellings is much sought after by the average mechanic. To acquire the art of making these drawings in a practical manner the learner must be shown how to proceed step by step. To look at a finished drawing the learner may get but a faint idea of the actual work and method of procedure. The full benefit of lessons in drawing can only be fully demonstrated by showing some of the work in different stages of completion, with a proper description of the method of proceeding from start to finish, and requiring the learner to exercise his skill and talent by duplicating the work. The art can never be successfully acquired without practice, and these lessons have been especially designed to supply the much needed practice.

It has been the practice of *Carpentry and Building* to present floor plans to a scale of 1-16 inch to the foot and elevations ⅛ inch to the foot. These scales are almost too fine for practical architectural drawing, yet for publication purposes they are more convenient than the larger scales. Should the floor plans and elevations accompanying these lessons appear to the 1-16 and ⅛ inch scales, it is specially requested that in the course of practice the draftsman draw the floor plans to ⅛ or ¼ scale and the elevations to ¼ scale, as by so doing he will acquire the art of drawing in a more thorough manner, because he will be compelled to make calculations for himself, and the art of calculating goes hand in hand with drawing. No one can successfully make drawings without calculations. The division and proportioning of the scale to different parts of the drawing are continually going on; therefore to enlarge the drawings to ¼-inch scale will in reality give the draftsman better practice than would be derived by copying them line for line by the smaller scales. After the drawings have been successfully made to the ¼-inch scale it would be well for the draftsman to

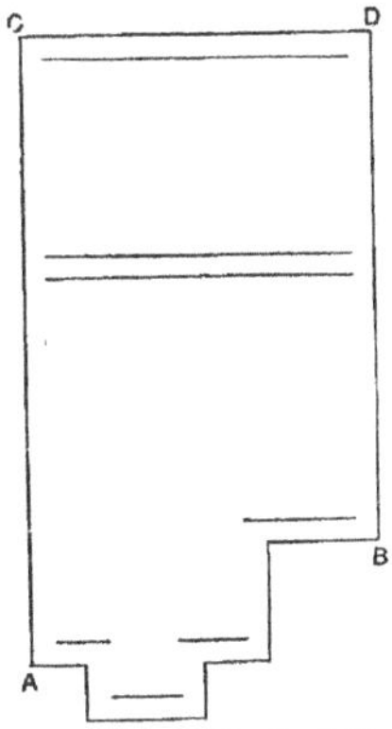

Fig. 4.—Method of Drawing Foundation Plan.—Scale, 1-16 Inch to the Foot.

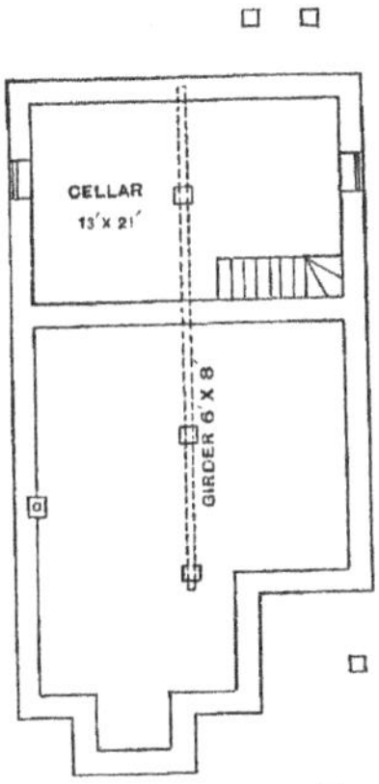

Fig. 5.—The Foundation Plan Completed.

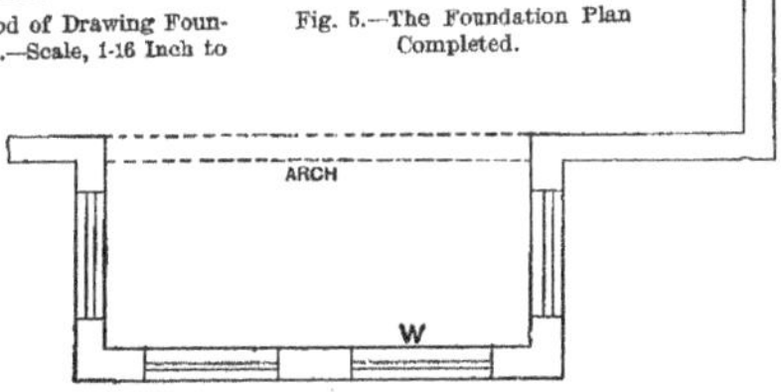

Fig. 6.—Portion of Front of Main Floor Plan, Showing Windows, Doors, etc.—Scale, ¼ Inch to the Foot.

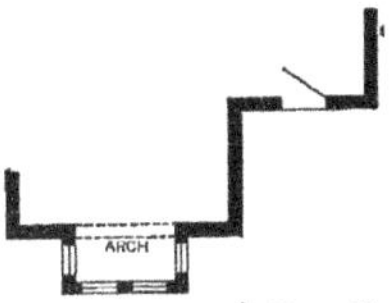

Fig. 7.—Front of Main Floor Plan Drawn to a Scale of 1-16 Inch to the Foot.

draw them to the ⅛-inch scale, as it will have a tendency to increase his accuracy in measuring and proportioning drawings, while at the same time it will extend his capabilities as a draftsman.

The first practice in drawing should be executed with a pencil, and it is quite as well not to attempt the use of ink until the learner has acquired a fair knowledge of pencil drawing. In the selection of pencils never choose soft ones, as they will not hold a point fit for drafting, and the crumbling of the lead is sure to work into the paper, giving the drawing a dirty, grimy appearance. A hard pencil is best, as with it fine clean lines can be made.

In making a drawing the outlines are the first consideration and should be made very light, so that in case an error occurs it can be easily erased. It is very likely, and more than probable, that beginners will make some mis calculations, and in drawing lines they are likely to draw past the stopping points or cross lines running at other angles and representing some other part of plan or design. It is necessary that the draftsman watch the work closely and acquire the art of starting and stopping at exactly the right point. This is one important feature in regard to drawing and a very essential one in regard to pen drawing. After the outlines have been correctly drawn the work can then be traced or redrawn, making the lines plain and distinct and completing the drawing throughout in a permanent manner. In regard to paper, it is unnecessary to have any particular kind, as any good quality with a hard, smooth surface will answer for ordinary practice, although drawing paper is preferable.

Having now given a general idea of the essential points in drafting we will proceed with some examples for practice, beginning with the foundation plan of a dwelling. The foundation plan is the plan of the cellar walls, foundation walls, piers, etc., and is usually very easily made. The starting point in outlining a drawing in nearly all cases should be at the front left hand corner, as at A, Fig. 4. The draftsman should work around the plan from left to right, although it is not necessary to adhere strictly to this rule. For example, the draftsman has several calculations to make in outlining the plan, particularly the front, which forms many angles. Starting at A, draw lines to B, representing the front, then from A draw the lines representing the left side, rear and right side as A C D B, which completes the outline of the foundation wall. The next step is to set off the thickness of the walls, draw the inside lines and partition walls of the cellar, if any are required.

The inside lines of Fig. 4 are so plain and easily understood that an explanation is unnecessary. In drawing the inside lines the main point to observe is to first draw all the lines representing the cross walls, making due allowance for thickness of walls, so that when the side wall lines are drawn the foundation walls will be complete. As a finished drawing does not show the method of proceeding as well as one partially finished, we will leave Fig. 4 in its present form showing the draftsman how to start, how to proceed and how to calculate the thickness of walls in order to locate the starting and stopping points in drawing the lines. Referring to Fig. 4, it will be seen that to complete the walls all that is necessary is to draw the lines representing the side walls. We will now proceed to finish by locating and setting off the chimney, piers, and girder under floor joists, cellar window frames, cellar stairs and piers for porches. This being done the drawing has the appearance of Fig. 5 and completes the foundation plan. Girders under floor joists are generally represented by dotted lines, as shown. Our floor plan and front elevation show two chimneys, but as one is intended to start from brackets on the partition wall only one will appear on foundation plan as drawn. Great care should be given to locating chimneys, so that they can be built plumb from their foundation to the finish. In many large cities there is an ordinance prohibiting the building of crooked chimneys in any form.

In order to explain a few points in regard to doors, windows, etc., we will present a portion of the front wall of the floor plan drawn to a scale of ¼ inch to the foot. Referring to Fig. 6, windows are usually designated by two lines drawn through the opening, as shown at W; outside doors are represented by one line, usually drawn on the outside wall, as shown at D. The way a door is intended to swing is designated by a line drawn to an angle, as shown. Inside doorways are represented by open spaces in the partitions with no lines across. The angle lines are frequently used when the draftsman wishes to specifically indicate which way the doors shall swing. Arches are indicated by dotted lines across the opening, as shown at the bay window, Fig. 6. They are also usually marked "arch," as shown.

To contrast the difference between an architect's working scale drawing and one as usually prepared for publication, the reader is referred to Fig. 7, which shows the front wall of the floor plan drawn to a scale of 1-16 inch to the foot.

DRAWING THE FLOOR PLANS.

As the draftsman has now been given an explanation of the primary principles and method of drawing plans, we will presume he is ready to advance another step in the art, and we will proceed to the floor plan in full. Referring to Fig. 8, first draw the outside wall line, set off the thickness of walls, locate the joining partition walls, and draw the lines, as shown. Having the wall lines drawn, the next step is to locate the doors, windows, chimneys, etc., which are represented by the short marks across the partitions. It will be noticed that in drawing the partition lines they have been drawn across the door openings, and also some of the lines cross at the junction of partitions where they ought not to. The advantage of drawing them this way in outlining will be plainly seen, for by having the outlines of all the partitions in sight the draftsman is better able to make calculations and locate doors, windows and chimneys in the most desirable manner. It should be remembered, however, that the outlines are to be drawn lightly with a pencil, then when we are ready to finish we can easily erase all superfluous lines, thus showing just where to start and stop when drawing permanently with the pencil or pen. This can be done a great deal easier and quicker than to make all the calculations before drawing the outlines; besides, there are often several doors and partitions in succession to cause confusion in locating exactly the points of starting and stopping.

We will now take Fig. 8, and erasing the lines across the door openings and joining partition walls will finish permanently, when it will have the appearance of Fig. 9. By comparing Fig. 8 with Fig. 9 the draftsman will be able to note the change that has taken place in the lines better than words can express it. In addition to the change made by the openings we have drawn a few more lines, finishing the stairs, sink, etc., and indicating the shelves in pantry and closets, making the plan complete.

In Fig. 10 is shown the method of drawing the roof plan. The inside lines represent the outside line of wall plate and should be drawn first. Next set off the width of the cornice and draw the outside lines. Locate the hips, valleys, ridges and chimneys and finish complete, as shown.

DRAWING THE ELEVATIONS.

In the usual course of architectural drawing the elevations come next, and very naturally they are the most complicated of all to make and appear as an insurmountable difficulty to

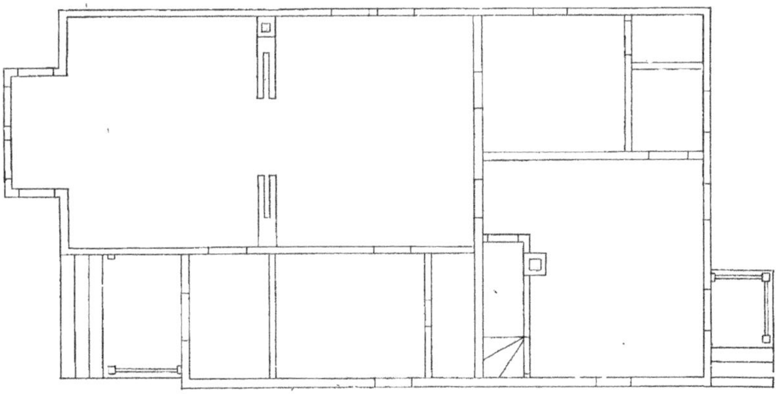

Fig. 8.—Method of Outlining the Floor Plan.—Scale, ⅛ Inch to the Foot.

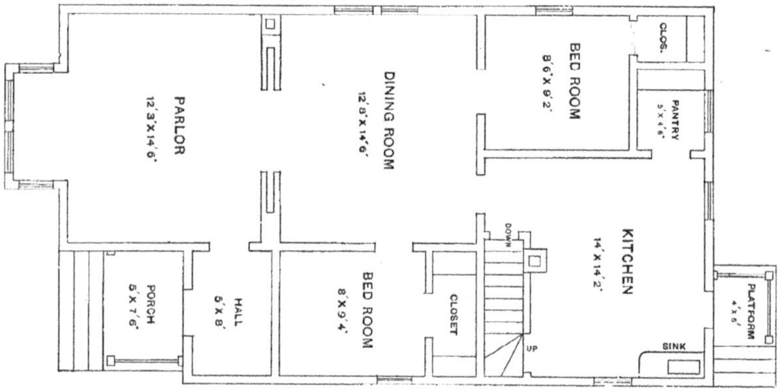

Fig. 9.—Appearance of Floor Plan When Completed.—Scale, 1/8 Inch to the Foot.

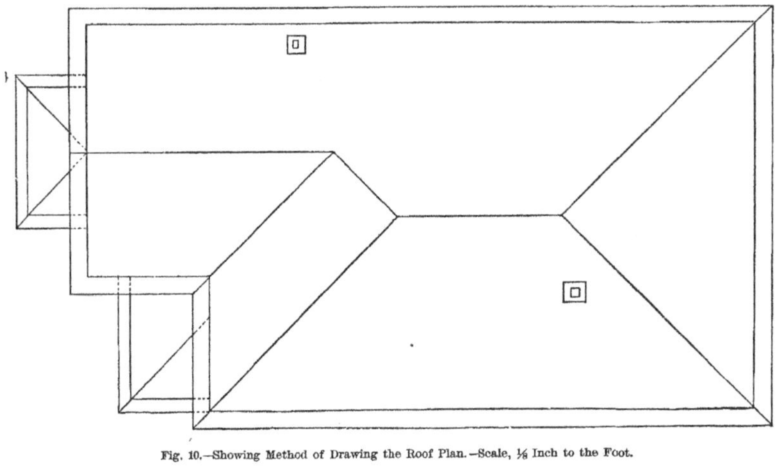

Fig. 10.—Showing Method of Drawing the Roof Plan.—Scale, ⅛ Inch to the Foot.

beginners in the practice of drawing. As a rule, all things come easy or comparatively so when we know how to proceed, and with proper instructions we hope those who desire will be able to readily master the difficulties usually met with. We will now start on the outlines of the front elevation, Fig. 11. Some may prefer to start from bottom of the foundation and build up, but we have found it more practical in drawing elevations to take the bottom line of the sill as a starting point from which to make the required calculations. It is always easier to calculate hights in an elevation from bottom of sill up, especially frame buildings, and for the hight of foundation it is no trouble to calculate from the bottom of the sill to any hight it is desired the foundation should show above the ground line. Hence we will take the bottom line of sill as a starting point and draw the line A B. Set off on line A B the width of front gable, as A C. Calculate the entire hight from bottom of sill to top of roof line, as shown by the scratch marks D, E and F, and draw the perpendicular lines A D, C E and B F, representing the main corners, but do not draw them quite to the points D, E and F, as these were calculated to the top of roof line, and the lines of the cornice must come below these points. It is easier and more practical to finish drawings from the top down, and when the lines representing the cornice are drawn they will intersect the perpendicular lines of the main corners in the elevation. This will be readily seen by comparing the unfinished work of Fig. 11 with the finished part of Fig. 12. It will be noticed that if the perpendicular lines were drawn full to the scratch marks we would have to cross these lines to finish the cornice, which, of course, would not be a proper thing to do. For convenience in outlining, the lines can be drawn lightly with a pencil and in full to the points D, E, F, and the part where the lines cross can be erased before finishing. This is a good way to do, as it shows the draftsman just where to start and stop when inking or tracing the drawing permanently. The points D, E, F are also the points from which to calculate the roof elevation, the amount for projection being added on at the bottom, as shown. After drawing the outline of roof locate and draw the outline of the chimneys. The latter can always be readily located from the floor plan. For example, the front chimney passes out the roof by the side of the left wall plate, while the kitchen chimney passes out the roof about 7 feet from the outside line of the right wall plate, which locates it where shown.

Chimneys that do not come out the highest portion of a roof should always be constructed somewhat taller than the ordinary chimney to insure them a good draft, and also for the protection of the roof against fire. It sometimes happens that roofs affect the draft of chimneys, hence the top of a chimney should be nearly as high as the roof through which it passes. Observe that the front chimney top is about on a level with the front gable roof and the kitchen chimney a little above the comb of the main roof. Chimneys which extend above the roof much more than the ordinary hight should have an anchor rod put in, as shown. The next step is to set off the width and hight of bay window from bottom of sill to lower edge of fascia. Set off width of crown molding and fascia, pitch of roof, and draw the lines as shown.

Next set off the outlines of the small window in the gable. In setting off the outlines of a window or door compute the extreme width and length, including casings, sill and cap, and draw the outside lines, as shown. It is more practical to finish a frame from the outside, working toward the center, than any other way. The drawing of doors and windows will be more fully illustrated and explained in the details.

The next step will be to outline the porch, which in all cases should be done before drawing the outlines of any doors or windows that appear under cover of the porch, because it very often happens that a portion of a door or window frame is partially concealed from view by some portion of the porch; hence the only proper method is to draw the porch first. Then, when the frames are drawn it will be plainly seen just what portion, if any, of them will be hid from view, and there will be no occasion for crossing lines. In most cases the bottom of the porch frieze hides the view of the head casings of frames that appear under cover of the porch. This feature is noticeable in almost every elevation of house design, and there seems to be no available remedy for it. The only plan would be to make the porches higher, and this cannot always be done for lack of space, and, again, extremely high porches look out of proportion and are equally as bad, if not worse, than hiding the head casings with the porch frieze. In the elevation the porch frieze covers the head casing, as will be plainly seen by referring to Fig. 12, which shows the work in a more finished state. The next step will be to set off the hight of foundation. Draw the ground line, space off the number of risers required and draw porch steps.

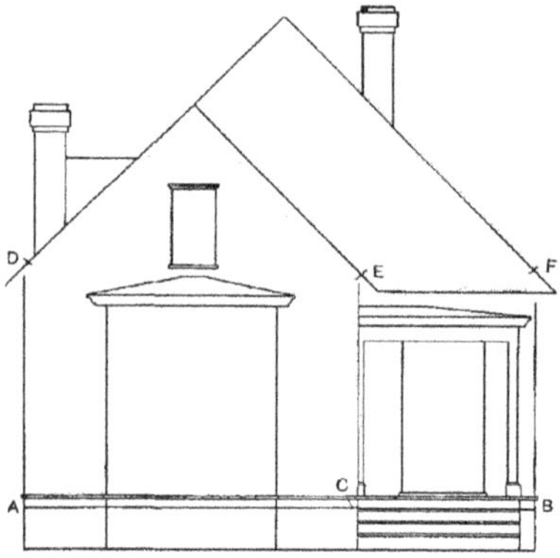

Fig. 11.—Method of Outlining Front Elevation.—Scale, ⅛ Inch to the Foot.

Fig. 12.—The Front Elevation Partially Finished.—Scale, ⅛ Inch to the Foot.

Fig. 13.—Appearance of Front Elevation When Completed.—Scale, ⅛ Inch to the Foot.

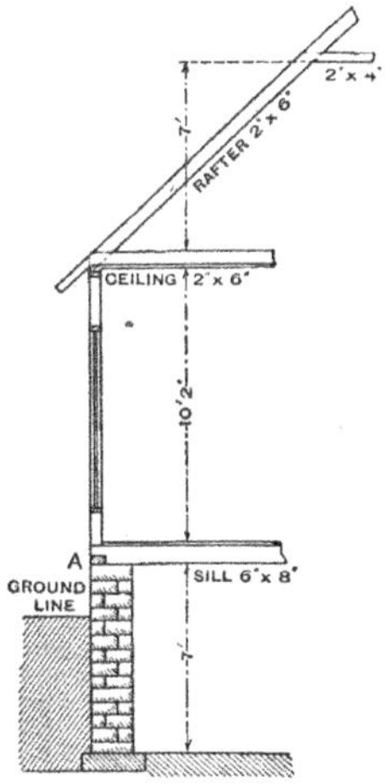

Fig. 14.—Sectional View of Framing and Foundation.—Scale, ⅛ Inch to the Foot.

The three parallel lines full length indicate the base and water table, which extend around the elevation on a line with the porch, as shown.

Trusting that a fair idea has now been given of the method of outlining and starting the elevation, we will next proceed to carry the work on to a more advanced stage of completion. Having made the outlines as in Fig. 11, proceed by drawing the gutter on the roof and finish off the hips. Next draw the lines representing the crown molding and fascia of the main cornice. Draw the gable ornaments, then the line representing bottom of frieze, and finish corner casings down to water table line, as shown. Next finish the attic window. The bay window will be next in order. Finish the lines of the crown molding, fascia and frieze. Draw the corner and middle casings down to water table line. Set off and draw window sills and finish by drawing window casings, sash and panels under windows.

Next we will take the porch and front door. Draw the lines representing the crown molding, frieze and columns, and finish the door frame, as shown. We will omit some of the details of the porch and bay window finish in this drawing, leaving the balance of the work for the drawing as it will appear when finished. It is better for beginners to see the work at different stages of completion, as it gives far better ideas of the method of proceeding.

The next step will be to show the work in a finished state by taking Fig. 12 as we have just left it and proceeding step by step to the finish, when it will have the appearance of Fig. 13. In finishing begin at the top and work down. Chimneys are usually the uppermost portions of drawings, although there are many exceptions in this particular. In this case finish the chimneys down to the roof line to represent the brick. It is not necessary to accurately scale every course of brick unless the chimney is of some special artistic design, in which case it would be necessary to have a detail drawing of it showing its entire construction. On common chimneys the size and total hight are observed in the drawing, the architect puts in a few lines to indicate the brick, and that usually finishes a chimney. We will now finish the roof. Begin at the top and make parallel lines to represent the roof. It is not necessary to scale the lines to represent the exact number of courses of shingles. It is usually understood and specified that the shingles shall be laid a stated amount to the weather. If there are any belt courses of ornamental shingles, then there should be a detail of the roof drawn to a suitable scale. Next finish bay window and porch roofs and finish down to the foundation, putting in the brackets and all ornamental finish, as shown. The siding comes next, and should be spaced according to the scale as nearly as practicable what it is to be laid to the weather. If possible always space to come out even at bottom and top of frames. In drawing the lines considerable care will have to be given the work in order not to draw across casing lines, porch columns and other parts of the finish. Drawing the siding lines is good practice; it aids very much to insure accuracy in starting and stopping at just the right points.

The next and last step is to finish the foundation down to the ground line. In this case we have drawn the foundation of stone, which is usually represented by somewhat irregular lines, as shown. We have now finished the front elevation, giving it the general appearance of Fig. 13.

We present in the next illustration, which is Fig. 14, a sectional view of the house, showing the size of timber, manner of framing, hight of cellar, hight of foundation above grade, hight of main story and attic. This drawing is so plain and easily understood that only a brief description is necessary. Taking A as a starting point, draw the bottom line of sill, then computing the hight of cellar and hight of frame to top of plate, draw the outside wall line from bottom of cellar to top of plate. Set off and draw outline of rafter from the plate. Next set off thickness of timbers and draw in the following order: Sill, outside wall studding, rafter, ceiling joist and collar beam. Set off the thickness of the cellar wall with proper footing, hight above ground line, and draw the lines as shown. Mark the hights of cellar, ceiling, attic and of headers to window openings, as shown in the outside wall. The front elevation was given at considerable length, showing every detail from start to finish, but we do not intend to go into the matter so thoroughly in regard to the side elevation. We have given the method of proceeding, and now present Fig. 15, showing the side elevation as it would appear in a finished state, hoping that those of the readers who may be considered as students in drafting will accept it as an example for practice and draw it to the best of their ability after the manner described in connection with the front elevation.

To aid the draftsman in making the side elevation we have just a little in the way of information to offer. After having the paper

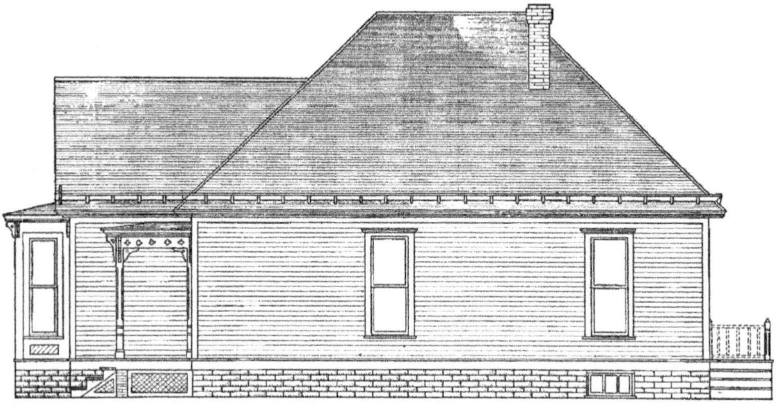

Fig. 15.—Appearance of Side (Right) Elevation Completed.—Scale, ⅛ Inch to the Foot.

fastened on the drawing board place the front elevation on it and to the right, so that the T-square will reach across both drawings. The result of this is that it establishes the hights for many parts of the side elevation and they can be transferred without making so many measurements. For example: The hights of porch, bay window, foundation, roof, chimneys and many other parts will all appear the same whether we look at the front elevation or the side elevation and consequently certain parts may be more accurately transferred in the manner above described.

DRAWING DETAILS OF A MODERATE COST COTTAGE.

The next lessons for practice will be with the details, which are portions of the elevations drawn to larger scales to aid the mechanic in carrying out the design, and are usually such parts as cannot be properly shown in the elevation. Some of these details will serve as the very best of lessons for practice, therefore it is to be hoped that the learner will give his best attention to the instructions that follow. We will now consider the details of a window frame for a two-light window, 24 x 36, casings to be 5 inches wide, with molded head casing and cap, scale ½ inch to the foot. Referring to Fig. 16, we first draw the outlines, then set off the thickness of the sill, width of molded cap and the casing lines.

The parallel line which extends around the inside of the frame next to the casing line represents the blind stop, and at the bottom of the frame it is the subsill. Next set off the thickness of sash, as shown. The two narrow inside lines represent the glass and putty lines. The division line of the sash extends across the frame from the blind stop lines. We have left Fig. 16 partially finished in order to show more clearly the manner of proceeding. It will be noticed that the perpendicular lines of the sash have been omitted. This has been done in order to show that after the outlines of the frame are made the next thing in order is to draw all the horizontal lines, or at least all that can be conveniently drawn, finishing the perpendicular lines last. Fig. 17 shows the frame as it appears in a finished state. It would be impossible to correctly represent all the lines of this drawing in an elevation drawn to ¼-inch scale, and in elevation drawings the fine lines are usually left out. The blind stop lines are frequently drawn, but not necessarily required in the elevations. In Fig. 17 we have drawn all the lines to show how a frame and window should appear when correctly completed.

Fig. 18 represents the front door and frame partially drawn. Fig. 19 shows the work finished. The drawings indicate so plainly the method of proceeding and being similar to what has already been described, with reference to Figs. 16 and 17, that further description is unnecessary.

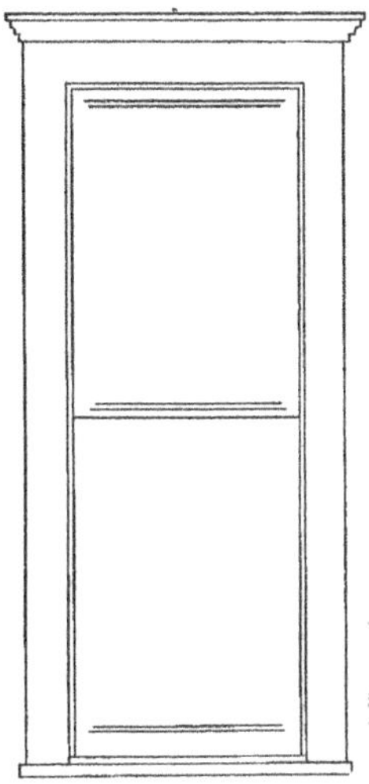

Fig. 16.—Outline Indicating Method of Drawing Window and Frame.—Scale, ½ Inch to the Foot.

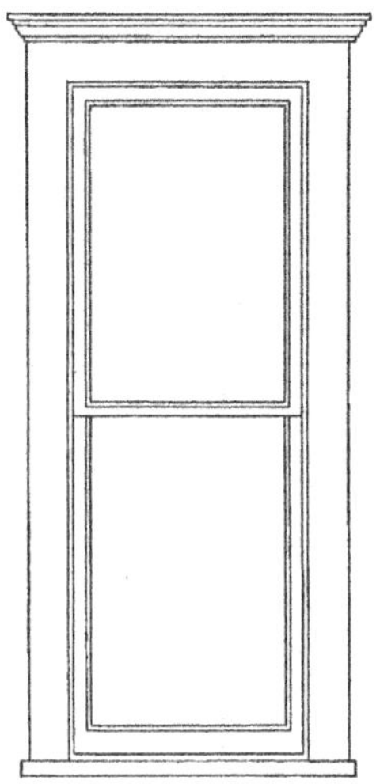

Fig. 17.—The Window and Frame Completed. —Scale, ½ Inch to the Foot.

Fig. 18.—Method of Drawing Front Door and Frame.

Fig. 19.—The Door and Frame Completed.—Scale, ½ Inch to the Foot.

A detail of the bay window drawn to a scale of 1/4 inch to the foot is shown in Fig. 20 of the engravings. The description and method of drawing the bay window was so thoroughly explained in connection with the front elevation that by comparing Fig. 20 with Figs. 11 and 12 the draftsman will readily understand the course to be pursued. A vertical sectional view drawn to 1/4-inch scale, which will serve as an example for practice, is shown in Fig. 21.

We will not give any prescribed method in drawing this figure, believing that it is just as well sometimes for the student to study the drawing and work out the method of procedure for himself. In explanation of the meaning of some of the lines we will say that the outside space represents the middle or corner casings, the second space the outside window casings, third space the sheeting, fourth space the studding, the fifth space the plastering, and the sixth space the inside casing. The sections representing the headers, head casings, sills, etc., are plain and easily understood. Fig. 22 represents a horizontal section, drawn to a scale of 1 inch to the foot, and shows very plainly the sections of the work through the bay window. In the engraving M C indicates the middle casing, O C the outside casing, B the blind stop, J the jamb and I C the inside casing. The other parts are fully explained by the wording. Fig. 23 is a detail of porch finish, drawn to a scale of 1 inch to the foot, and is so plain that little description will suffice. The ornaments in the frieze and corner bracket are simply open spaces. The particular point in this drawing to which it is desired to call the attention of the draftsman is the side and face views of the cornice bracket and the manner of showing the difference in appearance. Figs. 24 and 25 are details of the main cornice and of the work in the front gable, drawn to a scale of 1 inch to the foot, and need no further explanation. They will serve as examples for practice in drawing details.

A detail of the inside finish, drawn to a scale of 1 1/2 inches to the foot and showing the face and sections of base, plinth block, casing and corner block, is presented in Fig. 26. We have now completed the details of our house plan, but before concluding our lessons in drawing we wish to give a few examples for practice, to represent a class of work that has not been shown. Fig. 27 represents the shingling of a gable with round and octagon butt shingles, drawn to a scale of 1/2 inch to the foot. To execute such designs on a small scale is a very difficult task.

Fig. 20.—Detail of Bay Window.—Scale, ¼ Inch to the Foot.

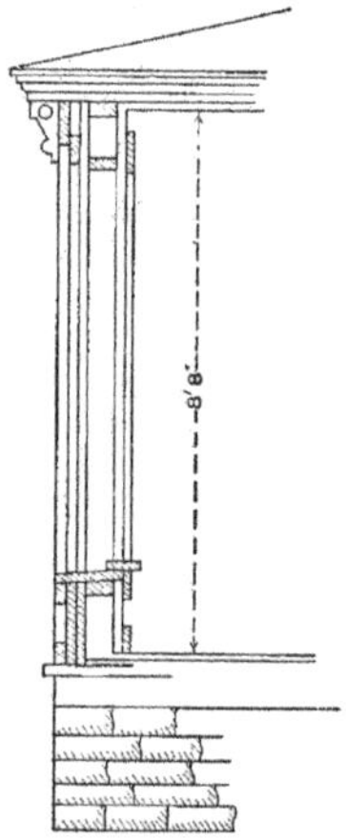

Fig. 21.—Vertical Section of Bay Window, Showing General Construction.—Scale, ¼ Inch to the Foot.

The method of drawing may be described as follows: After making the outlines of the finish down to the space required for shingles, space and draw lightly the lines representing the courses. Then, for the round butts, draw lines to use as centers in describing the circular lines, as shown by dotted lines in the engraving. For the octagon butts draw lines representing the depth of the octagon cut, as shown by dotted lines,

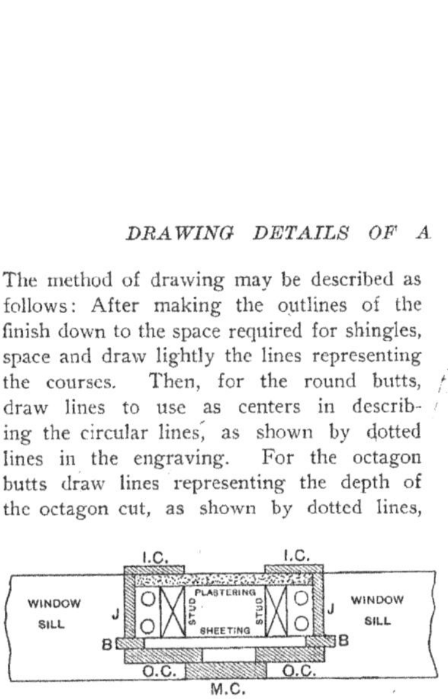

Fig. 22.—Horizontal Section of Bay Window, Showing Casings and General Construction.—Scale, 1 Inch to the Foot.

and finish as indicated, drawing the perpendicular lines last. After this has been done erase the superfluous pencil marks and trace the drawing permanently. A more expeditious and, perhaps, a better method to pursue with such work is to make patterns showing the profile of the bottom line of courses, as shown by Figs. 28 and 29. These patterns can be made of thin sheet brass, or even of cardboard, and will be found a great help in regard to speed as well as in keeping the work uniform. There are many small patterns of this kind that can be made to do good service in the way of helps and aids in drafting.

Fig. 30 shows an octagon plan of a bay window, and Fig. 31 the elevation, drawn to a scale of ¼ inch to the foot. Figs. 30 and 31 have been presented with a view of showing how to give a drawing the octagon appearance. This bay window is of the same general style as the square one previously shown, with but few exceptions. In this it will be seen that the window sill miters around the corners and the corner casings extend from frieze to sill and from sill to water table, instead of extending from frieze to water table, as shown in Fig. 20. It will be noticed that the side rail of the sash on the side windows does not show on the side next to the middle window. The reason for this is obvious when the plan is properly considered. Looking at an octagon bay window squarely from the front the casings of the frame hide this portion of the sash. In representing the brackets the draftsman has a regular picnic, and it is no small task to do the work properly. As will be seen by referring to the elevation, we have three views of the brackets. Directly in front we

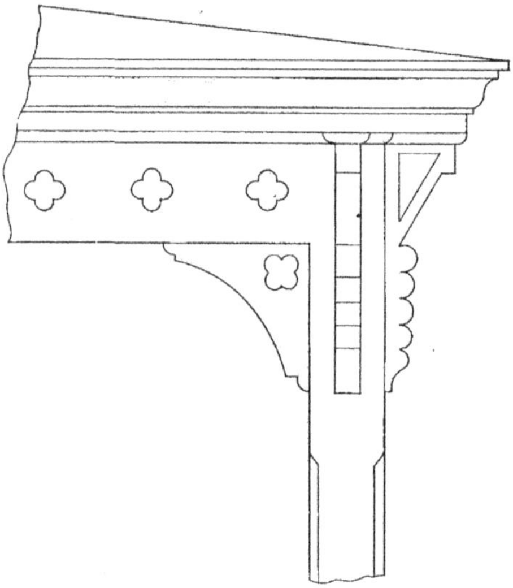

Fig. 23.—Detail of Porch Finish.—Scale, 1 Inch to the Foot.

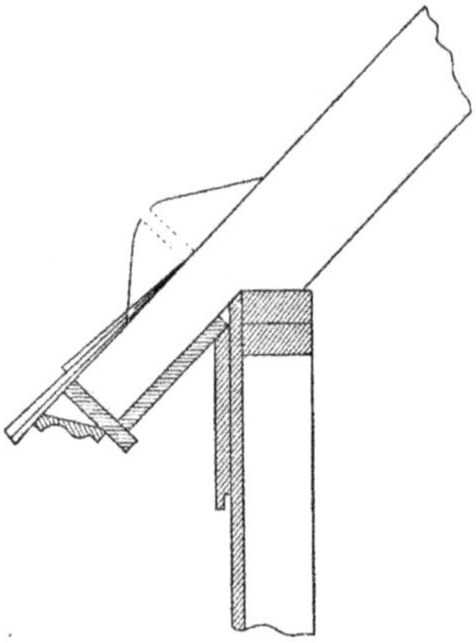

Fig. 24.—Detail of Main Cornice.—Scale, 1 Inch to the Foot.

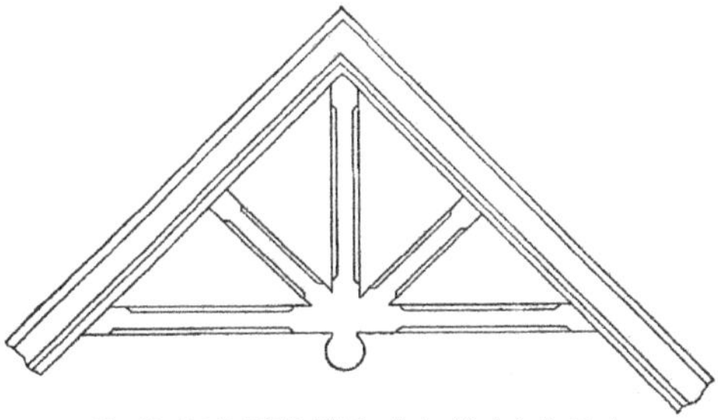

Fig. 25.—Detail of Gable Finish.—Scale, 1 Inch to the Foot.

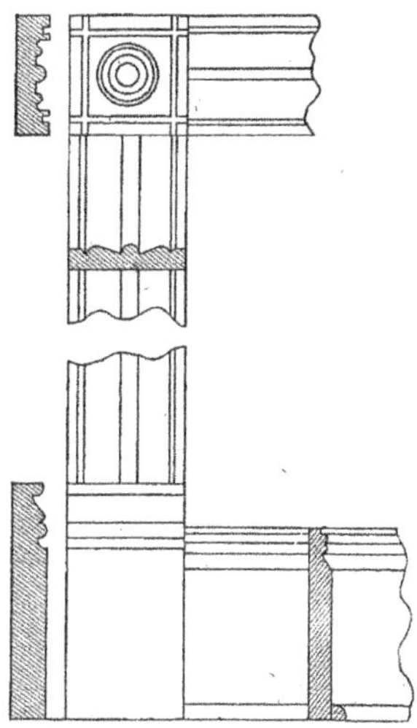

Fig. 26.—Detail of Inside Finish, Showing Faces and Sections of Base Board, Plinth Block, Casing and Corner Block.—Scale, 1½ Inches to the Foot.

get a face view, while next to either side we get kind of a half-face and half-side view. To the extreme right or left corner we would get a full side view of the bracket were it not for the fact that it is partially concealed by one of the side brackets. This drawing shows that there is a great study in architectural drawing and that much depends upon the judgment and skill of the draftsman. In fact, there are many things that come up in drafting that are very difficult to represent on paper. The draftsman has to form an idea of how the drawing should look to represent certain things; then study how to draw them, and lastly and most essentially, practice the art.

TRACING DRAWINGS.

We will now give a few instructions in regard to tracing drawings with ink. First, the draftsman wants a rule specially prepared for the work. Any common rule can be prepared for the purpose in a few minutes. Fig. 32 shows the shape of rule as used with the drawing pen, the same to be used with the beveled edge down. The idea of this is to prevent the rule touching the paper at the very edge of the ink line. When the rule and ink line touch the paper at exactly the same point, it is very difficult to move the rule without making blots. The rule slightly beveled and used in the manner above described is a sure preventive to blotting the work if anything like ordinary care is exercised.

In regard to the position of the pen, it should be held with the flat side of one of the nibs to the rule, and very nearly in a perpendicular position. The ink commonly employed is Higgins' American drawing ink, prepared ready for use. This ink dries quickly, the lines can be made very close together, and the rule moved around over the paper almost as fast as desired. The draftsman seldom has to wait more than a few seconds for the ink to dry.

In regard to filling the pen, there is a quill with each bottle of ink designed for the purpose. Our advice is to pay no attention to the quill method of filling the pen. It is too slow and bothersome. Dip the pen right in the ink, and then with a piece of cloth lightly wipe off any ink that remains on the outside of the nibs of the pen, as it would come in contact with the rule and otherwise might cause trouble. Always try the pen on a piece of paper before starting a line to see if it is working properly. On account of the quick drying qualities of the ink it frequently dries on the pen, thus stopping the flow of ink.

Fig. 27.—Method of Drawing Ornamental Shingles in Gables.—Scale, ½ Inch to the Foot.

Fig. 28.—Pattern for Shingle Work.

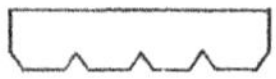

Fig. 29.—Another Shingle Pattern.

Fig. 32.—Shape of Rule for Use in Inking Drawings.

Figs. 30 and 31.—Plan and Elevation of Octagon Bay Window.—Scale, ¼ Inch to the Foot.

his is particularly bothersome, especially ı making very fine lines, and the nibs of the en have to be frequently wiped off and ljusted. Therefore we repeat, be sure the en is working right every time before you art a line, and then know just where you ant to stop the line when you start it. This about all there is of inking and drawing. emember to make all drawings lightly with ie pencil first, then ink the same lines over. any superfluous pencil marks remain they ın be easily erased with a rubber, which ill have no effect on the ink lines. To ink drawing on architects' tracing paper, place ie paper, which is transparent, over the pencil rawing, and the lines will show plain enough ı be easily traced.

We have now passed through a full de- ription of the manner and method of making chitectural drawings. The subject is one such wide range and the work that comes from time to time so varied that perhaps definite rules can be laid down that will meet the requirements in all cases. No two architects would be likely to pursue exactly the same course, yet they would arrive at the same results. To illustrate and describe all phases of the art would make a book seemingly without end. In these instructions in drawing we have given the draftsman the principles, method of starting and his course of procedure to the finish. By faithful practice and study of the art he will steadily improve, acquire the proper use of the tools and better judgment as to the methods of proceeding in difficult cases. These points must necessarily be left to the judgment of the draftsman as they come up from time to time. Experience and practice will prove of valuable service in the art, and gain for him a knowledge of drafting that can be acquired in no other way. Therefore let him bear in mind that success in the art depends largely upon a faithful adherence to the simple words of study and practice.

DRAWING A STONE AND FRAME DWELLING.

With a view to bringing out new phases of the subject, we next choose for consideration a large two-story stone and frame residence, with tower, octagon and round corners, as well as many other features calculated to produce a wide range of work and give the student the largest amount of study and practice possible to obtain from a single example. The basement and outside stone walls of the building are to be 18 inches thick up to the second floor, except the partitions of the basement, which may be 8-inch brick walls. All outside and partition walls of the frame part are to be of the common 4-inch studding, which, with the lath and plastering, makes about a 6-inch wall. The difference in the thickness of the stone walls and the frame partition walls causes a little study in the laying out of the rooms, in order to prevent the offsets in the walls making undesirable corners in the rooms. In preparing the floor plans shown in Figs. 33 and 34, the draftsman will see how easy it would be for this feature to show itself, and yet by a little study in the way of planning it will be noted how easily it has been avoided. The study of planning and designing goes hand in hand with the drawing, and if one is to draw practical plans every detail of the work must be thoroughly studied.

The first consideration in starting the work is the number of rooms to be provided, and the size or an approximation thereto, as it is not always possible to determine just how a plan will work out on the start. Draw the outlines in pencil so as to take in the desired number of rooms; then study how to divide them in the best possible manner for light, heat, size and general convenience. It will frequently happen that the general outline will need to be extended at some points and contracted or drawn in at other places; hence the outlines and general arrangement are the first points to consider. After these have been pencil-sketched and the plan has assumed a tangible form, it can then be per-

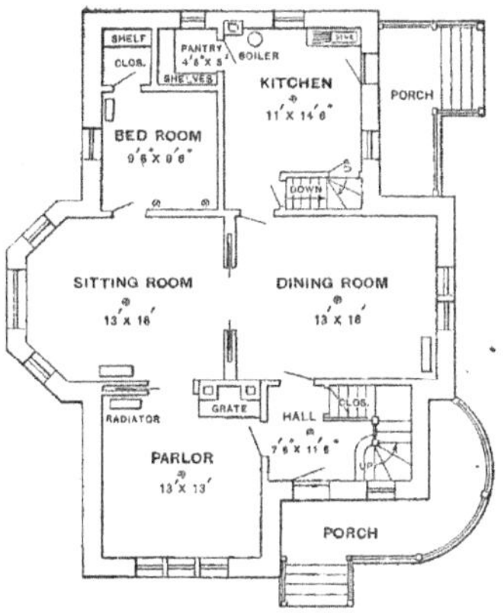

Fig. 33.—First Floor.

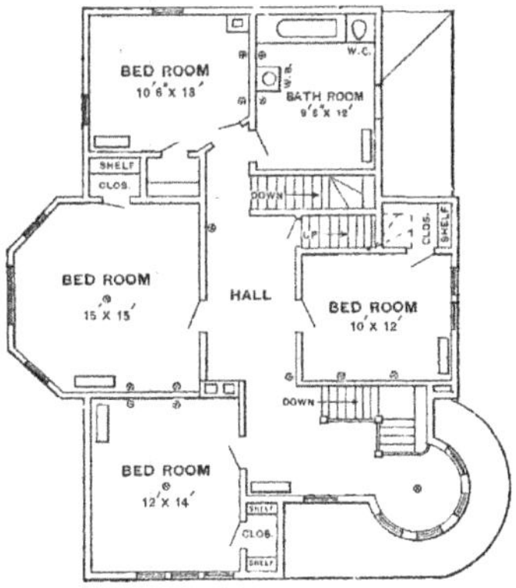

Fig. 34.—Second Floor.

Scale, 1-16 Inch to the Foot.

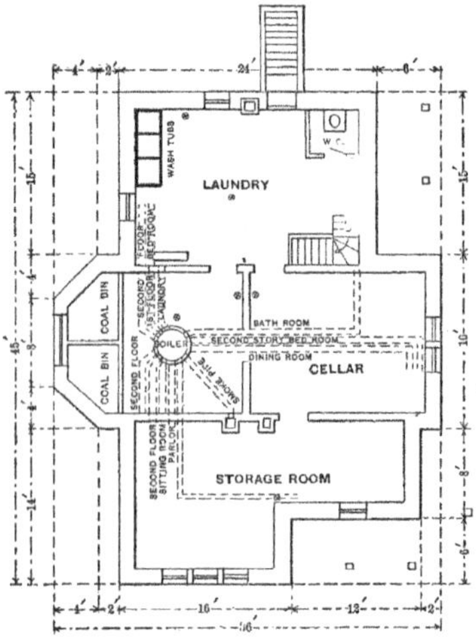

Fig. 35.—Foundation.

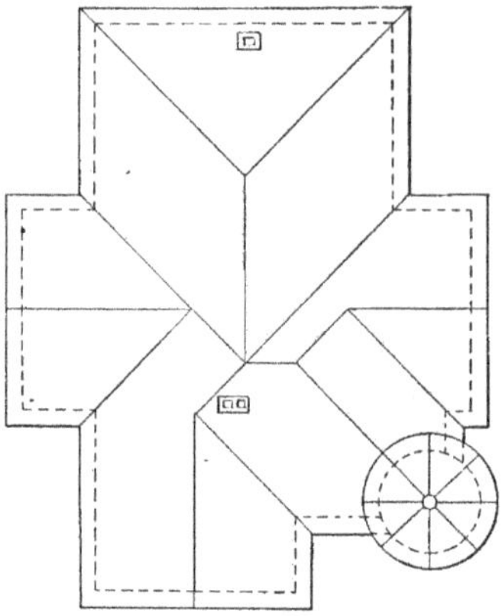

Fig. 36.—Roof Plan.

Scale, 1-16 Inch to the Foot.

manently drawn. The first-floor plan is usually the first consideration, and from it the other plans are correspondingly arranged and drawn.

The first-floor plan represented in Fig. 33 shows a few points not covered in previous work. For example, the burners for lighting are represented by a small circle and cross, indicating that the building is to be lighted by gas or electricity. The circles and crosses in the middle of the rooms are center lights from the ceilings, and those in the bedrooms are side lights from the partition, these being usually arranged about 5 feet apart to accommodate a dressing case, thus giving a light on either side. The boiler in the kitchen for supplying the house with hot water is shown near the chimney. In this connection attention is directed to the location of the bathroom on the second floor. It will be found directly over the kitchen, making the connections short and direct—a special feature of the plumbing that should always be considered. It lessens the labor and expense, while giving better service generally, to locate bathrooms as near as practicable to the water supply.

The house is arranged for steam or hot water heating, and the little rectangular spaces in the different rooms represent the location of the radiators. By comparing the first and second floor plans with the basement plan the general arrangement of the pipes will be seen. A special feature of this plan is in locating the radiators so that as many lines of pipe as possible may be carried parallel with each other to the different rooms. This results in a considerable saving of time and money, as well as obviates the necessity of a great deal of joist cutting, which is often the case where the pipes are widely scattered.

In the parlor, on the first-floor plan, is shown the method of representing a fire place and mantel. The small circles at

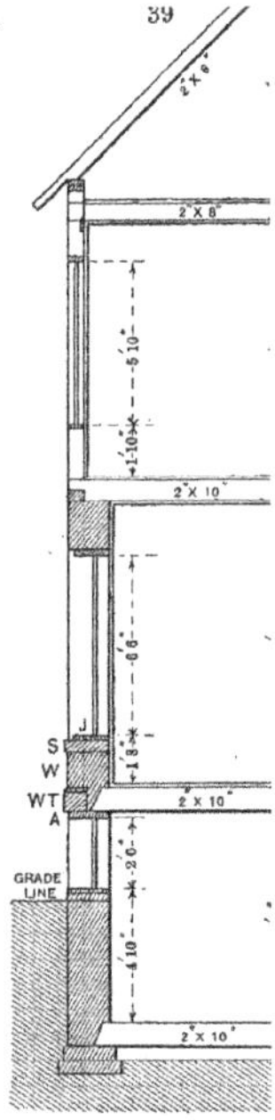

Fig. 37.—Section Showing I of Stories, Windows, Scale, ¼ Inch to the Foo

the outside of plan represent the down spouts or conductors for carrying the water from the gutters on the roof. The basement plan, Fig. 35, shows the general arrangement of the laundry, boiler room, coal bins, etc. The dotted lines outside of the plan show the outside dimensions of the foundation wall. The outside lines to the left and front show the size of the wall over all, which is 36 x 45 feet. The next set of lines shows the different divisions or the lengths of the wall from angle to angle. The total of the different divisions must equal the extreme length of wall on each side respectively. The drawing sets this forth so plainly that further description is unnecessary. It is obvious that by this method of drawing a plan a mistake in measurement is almost impossible.

The roof plan is shown in Fig. 36, the dotted lines indicating the wall plate and the outside lines the projection of the cornice. The other lines represent the hips, valleys and ridges, and are easily understood from the drawings. It will be observed that the octagon end of the left gable is finished square from the roof plates, having large brackets to support the projecting corners, thus giving a wider range and diversity of work, so that the draftsman may have a better chance to exercise his skill and talent in the art of drawing.

A sectional view showing the hights of stories, hights of windows above the floor, lengths of windows, etc., is presented in Fig. 37 of the engravings. The heavy shading or diagonal lines indicate the principal sections of the wall up to the top of the second story. Different sections should be drawn at varied angles, so as to show them more distinctly. This will be found especially valuable in representing members which join in close connection, as shown at A of the engraving. W T represents the water table, W the main wall, S sill, and J the jamb of frame and subsill. The hight and length of a window is represented by the distance between sill or subsill and head jamb, as shown. For example, take the first floor. We find the hight of story 10 feet between floor and ceiling; to the top of lower jamb or subsill, 1 foot 8 inches; hight of window, 6 feet 6 inches. In making the ordinary calculations for windows, 6 inches in length is usually allowed for the sash and 4 inches for width; 3 inches are allowed for bottom rail, 1 inch for meeting rail and 2 inches for top rail, making a total of 6 inches to be added to the length of glass. The side rails are 2 inches, making 4 inches in width to be added to the glass measure.

Fig. 38.—Front Elevation, Showing Method of Shading, etc.—Scale, ⅛ Inch to the Foot.

Thus a two-light 24 x 36 window will require an opening, exclusive of the jambs or frame, 2 feet 4 inches by 6 feet 6 inches. These figures are presented in order to give a little information as to the manner of calculation.

The frame part of the building starts from the second floor, and the general arrangement and construction will be readily understood from the drawing. In making the drawing, first draw the outside perpendicular wall line from bottom to top; then set off the thickness of walls, and proceed to draw the joist lines, floor and plaster lines, division lines of the different members and sections, and hights of windows and ceilings, leaving the section shading for the last. In figuring a drawing, as in setting off the hights, make the figures before inking the line, which will leave them plain and distinct.

The sectional drawing must correspond with the elevation as regards the hights of windows, sections, etc. It does not make any material difference which is prepared first if proper calculations are made. By making the sectional drawing first the proportioning of the windows to the hights of stories is much easier, because it shows the exact position of the windows as regards the hights on the outside, while the elevation shows the general arrangement and appearance on the outside.

Generally the elevation is drawn first and the section drawn to correspond, as it is not always easy to tell from the section just the hight at which a window will look best in the elevation, and a few inches difference on the inside is immaterial; but it must be borne in mind that the section and elevation must correspond, no matter which is made first. The best way for the draftsman is to make both the section and the elevation at the same time, on the same sheet and in the same line. Leave room to the left of the elevation for the sectional drawing and sketch the elevation; then sketch the sectional drawing and they will be in proper shape for comparison. Any changes in the planning and proportioning for improvement in general arrangement and appearance can be readily seen and made before the inking and finishing of the drawings.

The front elevation of the design is presented in Fig. 38, and as in a former portion of this work the manner of sketching and finishing an elevation step by step has been shown, we will not repeat the details at this time, but submit the elevation as an example for practice, calling attention only to the features worthy of note. Working drawings do not necessarily require anything more than plain lines; but as drawings for publication are frequently embellished with more or less shading to give the design an artistic appearance, attention will be directed to a few points on shading. Shading is not one of the necessary requirements, but by proper shading a more pleasing effect to the eye can be produced, and the draftsman who can display the most artistic skill, combined with other qualifications, is most likely to meet with success. Therefore it is hoped that the few ideas which may be presented will prove but the starting of work which the readers will carry out much better than here indicated. We will give the ideas, but the manner of best presenting them in drawing can only be acquired by study and practice.

The starting point of the elevation is taken on the line A B, which is the bottom line of the stone base or water table. The stone below this line may be considered the foundation walls, and the shading is represented by somewhat irregular diagonal lines—a common way of representing stone foundations. The block stone above the water table is shaded with diagonal parallel lines, every stone not being shaded, but taking them in an order that prominently distinguishes the corners and openings of the building and obviating the plainness in appearance which would otherwise result. The slightly irregular curved lines on the corners shown from course to course of the stone indicate that the stone are rock faced. If the stone were square faced the corners would be represented by a straight line.

In the windows the deep black shading represents the glass, while the scallops extending through the windows in a somewhat diagonal manner indicate lace curtains. The two windows to the left in the drawing do not show the right side sash rail, for the reason that this portion of the wall is an octagon corner, and in a direct front view this portion of the sash is hidden. The roof, however, on this octagon end has been framed similar to a square or plain gable roof, and the projecting corners are supported by large curved brackets, as shown in the elevation. It will be noticed that the roof of the tower is cross lined and appears finer toward the outlines, so as to indicate circular construction. Having now briefly referred to a few points on shading, it is suggested that the draftsman increase his knowledge in this particular direction by closely studying architectural designs as presented in first-class architectural journals.

Referring now to Fig. 39, the right side

Fig. 39.—Side (Right) Elevation, Showing Parts of the Work in Different Stages of Completion.—Scale, ⅛ Inch to the Foot.

elevation, we will show the several parts of the work in different stages of completion, believing this to be the best method of clearly indicating the plan of procedure. As before stated, the design should be first sketched in pencil, in doing which it is often necessary or convenient, at least, to make some superfluous lines or marks to aid in setting off different parts of the work, but which are not required in the finished drawing. These lines are often necessary in making calculations, and in order to show their usefulness and distinction all such will be represented as dotted lines in Fig. 39.

It should be borne in mind that all the work is first sketched in pencil, while in the completed drawing only the lines required are inked, after which the superfluous lines, represented as dotted lines, may be erased. By closely following the sketch step by step the draftsman will soon become familiar with this work, as a portion of every part, both in a finished and unfinished state, will be shown. The starting point is from the line A B, the bottom line of the stone water table. After this has been drawn set off the hight of the foundation above grade and draw the ground line, as shown. Next set off the angles shown in the plan on the line A B; compute the hight at the top of rafter or roof at the wall plates and draw the plumb lines of the corners. As the cornice extends below the top of these lines, the portion which extends through the cornice lines is indicated by dotted lines. The tops of these lines, as shown by C, D, E, F, G, are the points from which to make the calculations for the elevation of the roof, and after establishing these points compute the hights of roof at the essential points and draw the general profile of the roof in outline, taking the gables, ridges, hips, valleys and cornice lines. Next outline the finials, cresting, chimneys and gutters. The dotted plumb line showing center of tower and the portion of dotted hip line joining at D plainly indicate that it is necessary to locate this point in order to correctly represent the roof. By comparing Fig. 39 with the first-floor plan it will be seen that the dotted plumb line virtually represents one of the main corners of the plan, but the tower being built out on this corner from the second floor would of course hide the line from view, consequently it should not appear in the finished work. Attention has been called to this simply as one case perhaps out of hundreds where the drawing of temporary lines will aid in making calculations and completing the permanent drawing. After these temporary lines have served their purpose they should be erased before inking or finishing the work. The dotted lines shown in the drawing of the front chimney and finial on the tower indicate their usefulness again in shaping and setting off the different parts of the work.

The next step is to locate and outline the upper windows. First find the proper hight for bottom of sill and draw the line which will represent the bottom line of frames. Locate the windows on this line and set off the thickness of the sill. Next set off and draw the size of the opening, allowing for thickness of sub-sill in the hight. Set off the width of casings, cap, etc., and draw the lines finishing the frame, as shown in the upper right corner of the elevation. The manner of drawing the sash in the frame is plainly shown by comparing the finished window with the unfinished one in the double frame at the left. This method of setting off and drawing frames will be found easier and better than the one set forth in an earlier part of this work, as the draftsman is not obliged to make calculations so far in advance of the work. The method referred to is the outlining and drawing of frames, and shown in Fig. 11 of a previous article.

The next step in the progress of the work is the outlining of the porches, second-story base, first-story frames and so on down to the ground line.

Having the building now thoroughly outlined, we will proceed to finish portions of the different parts, so that the learner can see more readily just how to proceed. We will not go through with the finishing step by step, as we believe that the work, as shown in Fig. 39, is presented so plainly that further description is unnecessary. It will be sufficient to say that in finishing begin at the top and work down, which avoids to a great extent rubbing the tools and hands over the finished work, and will aid very much to keep the drawings looking neat and clean.

A LESSON IN OUTLINING.

For a lesson in outlining which will present a few new features in drawing the attention of the student is directed to Fig. 40, representing in outline the left side elevation. It will be noticed that this view of the plan presents an octagon end from the ground line to the eaves, finishing with a square gable, the projecting corners being supported by large brackets. This is a form frequently met with in practice, and will therefore serve as a valuable lesson to those who wish to make a special study of the art of drawing. Beginners are liable to grasp the idea that an elevation showing an octagon design should be represented by drawing the side lines of the octagon on a slight angle, as shown by the dotted lines next to the ground line. If this were the case, then all the parallel lines on these sides would necessarily have to be drawn on the same angle from the ground line to the starting of the roof. Such a course as this would give the drawing a rather crooked appearance. It should be remembered that all horizontal lines shown in a direct face view of an octagon elevation are to be drawn straight across all sides, as shown. The miter lines shown so distinctly in the perpendicular lines representing the corners plainly indicate the portion of the elevation having the octagon form.

We do not consider it necessary to give more than a brief description of outlining this elevation, believing it best to leave the greater portion for the study and practice of the learner. First draw the base line, set off the corners, and then draw the perpendicular lines of the corners. Calculate the hight to the eaves and draw the bottom line of roof—not the cornice line, but the shingle line, for example. Next compute the hights of roof, gables, etc., and draw the outline of the roof. Locate and sketch the chimneys, then the crestings, finials, etc. It will be noticed that a comparatively small portion of the front chimney is visible in this elevation, for the reason that as viewed from the left side the roof hides a portion of the chimney. The same is also true of the tower, only a small portion of the top and the finial being visible from the left side elevation.

Fig. 40.—Side (Left) Elevation in General Outline.—Scale, ⅛ Inch to the Foot.

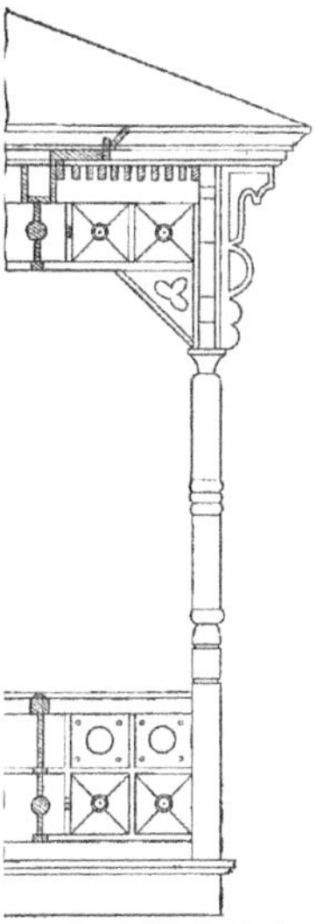

'ig. 41.—Detail of Porch, Showing Method of Drawing the Different Members.—Scale, ⅜ Inch to the Foot.

These points serve to show that the plan and corresponding elevations have to be carefully studied and watched during the entire progress of the work. In the sketch is shown a portion of the cresting, finished, in order that the difference in the views from the right and left sides of the same may be distinctly seen, as in one view the tower appears in front of the cresting and the other shows what would be the exposed portion back of the cresting. These points will be plainly seen by comparing Figs. 39 and 40. In outlining the gutter the lower portion is represented by dotted lines, because in the finishing the short perpendicular lines representing the small brackets should be drawn before drawing the horizontal lines between the brackets. This is plainly shown in Fig. 39.

After outlining the roof, the next in order will be the cornice, such as the molding, frieze, etc. Some advantage can be taken by drawing the parallel cornice lines at the same time the eave line is drawn, as they are in close connection, and all that is necessary is to make proper calculations in doing the work. The next step is to draw the base and water table lines which mark the division of the two stories so distinctly in this drawing. Next calculate the hight of windows to the bottom of sills and draw the lines which are to represent the bottom line of windows. From these lines set off and draw outlines of windows, as shown.

We think now that the details of outlining have been made sufficiently plain to enable any one to go to work understandingly and complete the drawing. We would recommend that as a lesson for practice nothing could be better than for the learner to complete the unfinished work shown in Figs. 39 and 40 in a manner similar to the work shown by Fig. 38, using a scale of not less than ¼ inch to the foot. Fig. 41 shows a portion of the porch finished in detail to the scale of ⅜ inch to the foot. First draw the base and floor lines, then two perpendicular lines representing the column. Set off the hight from floor to bottom of frieze and draw the bottom line of porch frieze finish, which will give good starting points for all future calculations. From the bottom line of frieze finish the different parts may be readily set off to top of roof. From the floor set off the hight of railing and draw horizontal lines first, spacing and filling in as shown. The turned portion of porch column may be easily drawn by spacing and setting off the hights of the different members, as shown. The sectional parts show the general construction of the work. The spindles are ball turned and set between square upright pieces, as shown. For general practice we would suggest that such details be made on a scale of ½ inch to 1 inch to the foot. We have now passed through the several parts of the work of making a set of plans which bring to notice a very wide range of work in the art of drawing, and if the learner has thoroughly mastered the work thus far he is qualified for further advancement.

DESIGN FOR STORE FRONT.

It is hardly necessary to go into all the details of outlining the elevation, section and plan of the store front, as we believe sufficient has been said upon the subject to enable the learner to study the work which will now be presented and to master it without other suggestions than a few brief instructions. The knowledge gained by the practice and experience of working out the several parts of the drawing for oneself will be far more beneficial to the learner than to have mentioned every little detail, many of which have been thoroughly explained in connection with other drawings. We will, for the most part, take up such portions of the work as possess new features. A special feature to which we wish to call attention is the corresponding lines of width and hights in the plan of the front and sectional view as compared with the elevation. The drawing, Fig. 42, is a combination of the plan of front, sectional view of the walls, etc., and the elevation, and it shows plainer than words can describe the relation one part bears to the other. It shows very plainly how to draw the elevation in accordance with the plan of front and to carry up with the greatest degree of accuracy and convenience the section corresponding with the hights shown in the elevation. By this method mistakes and discrepancies in the different parts closely connected are avoided. In order that the learner may start aright, it may be stated that the sidewalk line in the drawing is the most favorable starting point, and it will be found easy to make calculations from this line in any direction necessary to complete the work. The sidewalk line is the base line of the elevation, therefore set off on this line the width of the building and square down sufficiently to draw the plan of the front. Draw the outline of the front and the thickness of the wall; then set off the doors, windows, columns, etc., as shown in the plan. The dotted plumb lines show how to carry up the front in accordance with the plan. The location of the second-story windows, as a matter of fact, necessarily has to be in accordance with the front of the second-floor plan. The circular dotted line on the left corner of

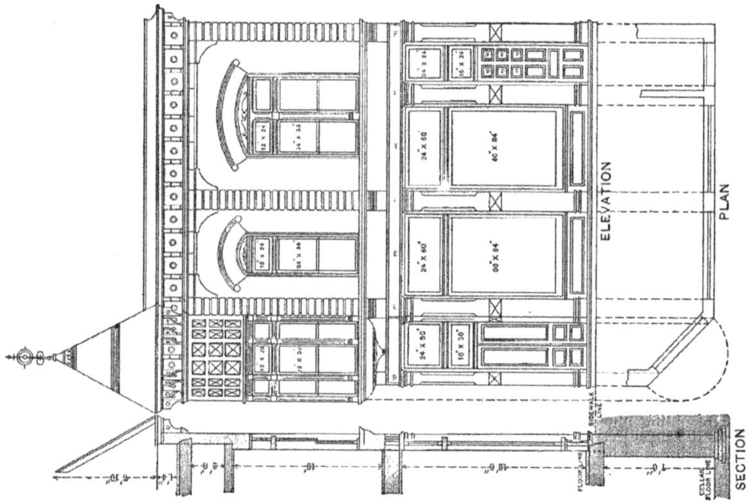

Fig. 42.—Showing the Relation of the Plan, Elevation and Section of Store Front to Each Other and the Method of Drawing.—Scale, ⅛ Inch to the Foot.

the plan of the front has been drawn to show how to get the required projection of the oriel window, built out from the second floor. This is plainly indicated by the long dotted plumb line. The starting of the support to the pro-

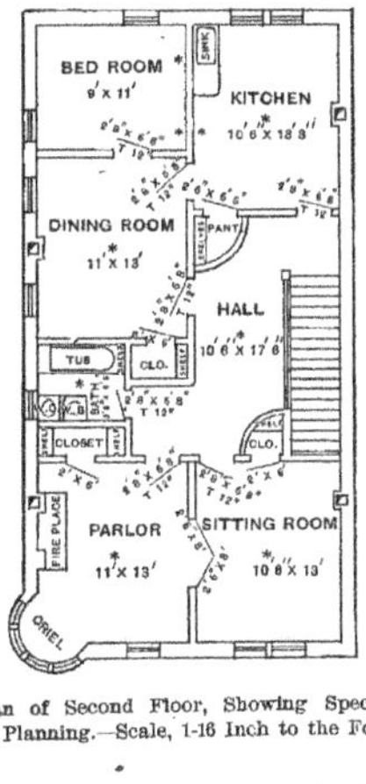

Fig. 43.—Plan of Second Floor, Showing Special Features of Planning.—Scale, 1-16 Inch to the Foot.

jecting window is from a point plumb over the center of the octagon corner in the plan. This, of course, makes the outside scroll appear more elongated than the inside scroll. The window being round and starting on the side of an octagon corner accounts for this appearance, and a little study of the plan and elevation will make this point clear to the mind.

By comparing the elevation with Fig. 43, the second-floor plan, it will be observed that either a direct front view or a direct side view of the plan would show us three windows or part of three windows in either a front or side elevation. In the front elevation there is in reality more of the outside window shown than is in strict accord with the second-floor plan. This point will serve to impress upon the mind of the draftsman that it will often require his closest attention in comparing drawings and his best judgment in executing the work to keep the proportions in accordance with the point of view taken of the plan and elevation. Of the section showing hights of ceilings, etc., very little need be said. It should be sketched with the elevation as a guide to correspond with hights of windows, floors, etc. The first-floor hight shows a section of the store front sash. A section in line of the doors would be slightly different, as double transoms have been provided in order to lessen the hight of the doors, as shown.

This store front is designed with heavy cut stone columns from base to the I-beam. From the top of the I-beam the columns are continued with block stone to the cornice line and a stone coping is run across at the bottom of the second-story windows. The oriel window, tower and cornice is of frame construction, although the same design can be executed in sheet metal on a wooden frame work. The body of the work is to be of brick, as are also the arches over the windows. The caps to the windows are to be galvanized iron. There are no brick lines drawn in the elevation, as these will be left for the practice of those who wish to still further carry out the design and improve upon it.

HINTS ON PLANNING.

We will now give a few hints in regard to planning, a feature which should also command the attention of good draftsmen. Small and irregular shaped rooms should be avoided. All rooms should have square corners if possible; projecting corners and recesses should not appear in a plan unless caused by a chimney or a bay window. There is nothing nicer than plain, square cornered rooms, especially when it comes to the carpeting and furnishing. Octagon corners should generally be avoided; also three-cornered closets across the corners of rooms. A much better way for such closets is to allow a little more room in a hall or some other part and build the closet out from the room, as shown on the floor plan, Fig. 43. This circular form gives a much better shaped closet by avoiding the sharp corners, giving more room and leaving the rooms with all square corners. There is also plenty of room left in the hall. Long and narrow halls should be avoided, also steep and crooked stairs. The architect and designer should plan to reach as nearly as possible all the rooms on each floor of the main hall and to use all the room to the best possible advantage.

In the floor plan is shown the method of marking where transom frames are used in the interior; thus 2′ 8″ x 6′ 8″ T 12″ means that the transom to be used should be 12 inches in hight. The elevation shows the method usually employed in marking size of windows and the figures give the glass measure.

STAIR WORK.

We will next touch upon the subject of planning, laying out and drawing stairs. Everybody knows that many a time too little room is left for the stairs, and the result is steep and awkward stairways in what are supposed to be the better class of houses. Much of this is due to ignorance or carelessness on the part of the designer to make a few figures. It is too often left in this way: "I guess so much for stairs will be plenty." The best way is to figure it and make sure that there is plenty of room. For the benefit of those who are young in the work of planning we will show by sketches how to determine the amount of room required, and also give an easy method by which to solve the problem by figures.

Referring to Fig. 44, which represents a straight flight of stairs, we find the rise as given from floor to floor to be 10 feet. By computing the risers at 8 inches we find that it takes 15 to reach the top floor. It will be seen that we calculate space for only 14 steps, as the fifteenth lands on the floor. Next we calculate the run required for 14 steps. These we have estimated at 10 inches each, making the run 11 feet 8 inches, as shown. The rise and run of stairs are always taken as laid out on the string board, the projection of the step for nosing and molding, as shown at the bottom step, not being counted. From the top of the third riser to the ceiling is 7 feet. Plumb over the third riser, and facing it, we have located the header, giving 7 feet head room for the stairs. From this header we count the number of steps back, allow for thickness of riser, an inch or so for work room, and we have the distance between headers, or the length of the well hole, as it is sometimes called. In this case it is 10 feet 3 inches. The sketch shows very plainly how to figure the stairs with a certainty as to the room required, and the same general plan will hold good in any case. The sketch shows an easy pitch. Stairs are frequently run much steeper, but for good work we advise plenty of room for easy stairs. If it is desired to figure it without a draft, commence at the foot of the stairs, proceeding as follows: First

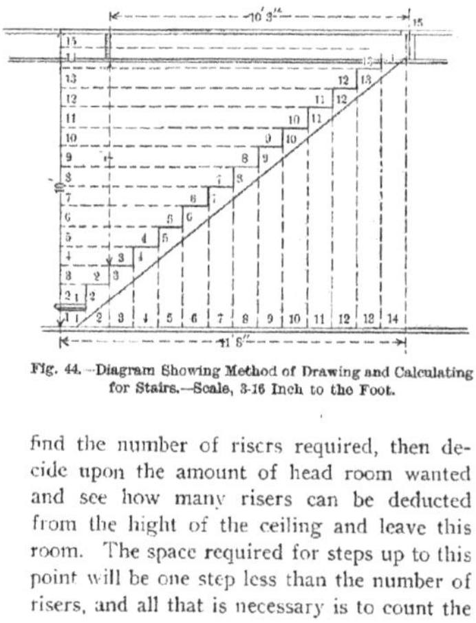

Fig. 44.—Diagram Showing Method of Drawing and Calculating for Stairs.—Scale, 3-16 Inch to the Foot.

find the number of risers required, then decide upon the amount of head room wanted and see how many risers can be deducted from the hight of the ceiling and leave this room. The space required for steps up to this point will be one step less than the number of risers, and all that is necessary is to count the steps, lay them off on the floor and start the header plumb over the back edge of the last step, or as nearly so as practical. Fig. 45 shows the method of laying out and drawing the plan of winding and platform stairs, with a portion of the framing for the floor joists, headers, etc. To find the exact floor space required for the stairs, lay off the run of each step according to actual scale measurement, as shown. It

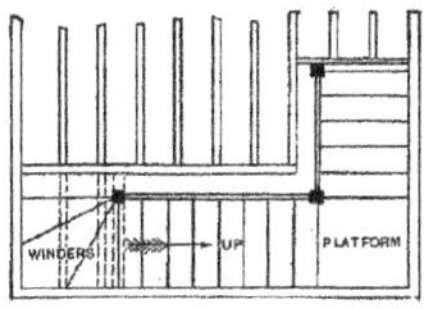

Fig. 45.—Method of Drawing, Calculating and Framing for Winding Stairs.—Scale, 1-16 Inch to the Foot.

is customary to place three steps as winders in stairs of the average width; sometimes four steps are placed in the winders, but we would not recommend more than three and never more than four steps in the winders. The space required for the winders is usually taken as a square, the sides of which equal the width of the stairs. For example, if the stairs are 3½ feet wide, the space required for the winders will be 3½ feet square. The same rule holds good in regard to platforms. As regards the framing for the headers, it is safe to say that a header could be placed flush with the side of the square and third winder, and leave plenty of head room in ordinary dwellings having 9 to 10 foot ceilings, as shown by dotted lines extending across flush with the first post. This would leave room enough on the floor above for a small closet. If this was not desired, or if for any reason this arrangement should not give head room enough, the short header and joists indicated by the dotted lines may be omitted and the frame work put in as shown. From 6 to 10 inches is usually allowed between the string board and the headers for stairs with a rail, or "open stairs," as they are frequently called. At the landing it is only necessary to allow 2 or 3 inches between the header and the last riser. This sketch is given to impress upon the mind of the draftsman the importance of good calculation in connection with drawing plans, and that a scale drawing is the best and most effective way to show the calculations.

FRAMING PLANS AND ELEVATIONS.

We will now turn to a few drawings showing the method of indicating framing plans. Fig. 46 represents the plan of a farm barn with large floor space, driveway running through the building, stable room for five horses, granary and passageways. This plan is well arranged for an addition or an annex to the right, such as a cow stable, for example, which would be handy to both hay and grain, and would be considerably removed from the horse stable, which is a feature generally desired. It is hardly necessary to show in detail the manner of drawing the sketches presented on this subject. The drawings are plain and easily understood and will serve as good examples for practice, particularly Fig. 47, which shows just where to start and stop the pen without crossing lines which should not be crossed in representing the work properly. This figure shows the general framing plan of the sills and floor joists. It is desirable in stables to have the floor run lengthwise behind the horses, with a double floor in the stalls, the top floor running lengthwise of them. In the driveway it is desirable for the floor to run crosswise, as it is not as slippery to the horses' feet when pulling in loads. It also makes a stronger floor. This we trust will be sufficient reason for the plan of the joists presented in Fig. 47. The plan shows the joists resting on top of the sills and girders, which is a better and stronger method than framing them in, besides being easier and

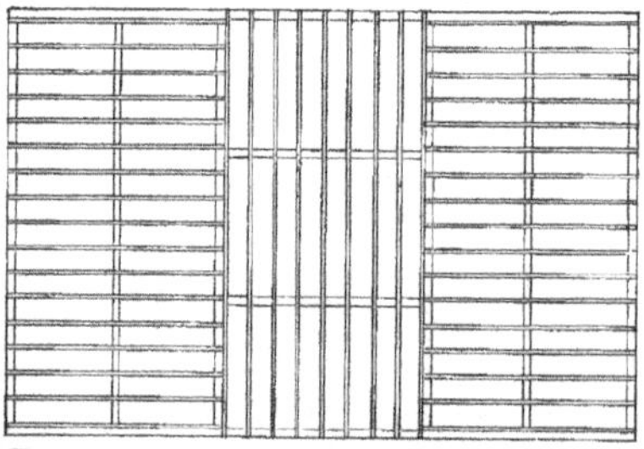

Fig. 47.—Method of Drawing Floor Framing Plan.—Scale, 1-16 Inch to the Foot.

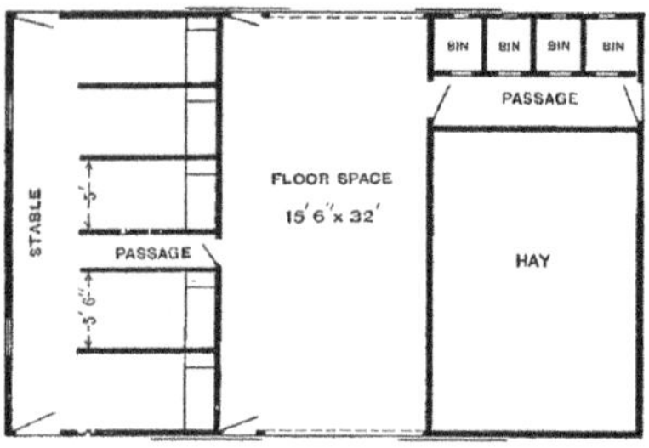

Fig. 46.—Floor Plan of Barn.—Scale, 1-16 Inch to the Foot.

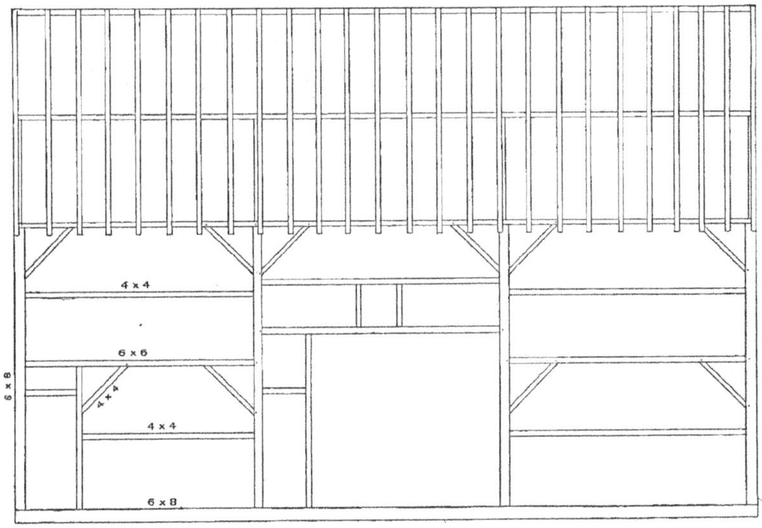

Fig. 48.–Method of Drawing the Right Side Elevation of Frame.–Scale, ⅛ Inch to the Foot.

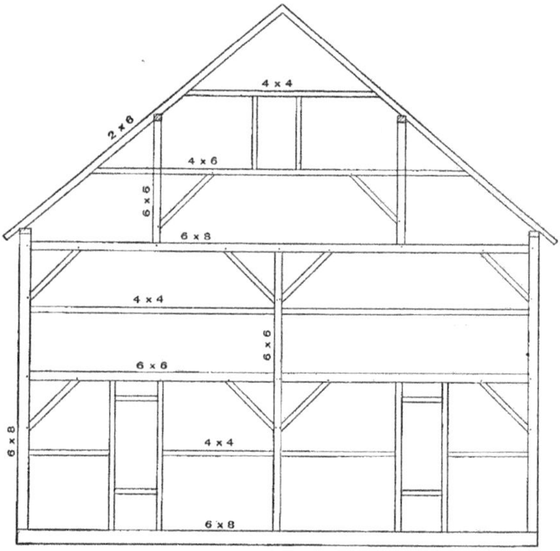

Fig. 49.- Method of Framing Left End Bent.—Scale, ⅛ Inch to the Foot.

quicker. A little study of the framing plan, Fig. 47, will be sufficient to show the draftsman just how to proceed to draw a similar plan.

The framing of the right side elevation of the building is shown in Fig. 48 of the illustrations. The particular part to which attention is invited is the starting and stopping of the lines at just the right points in drawing the plates and rafters. As previously stated, the first thing is to pencil sketch all work; then with well guarded movements one can tell just where to start and stop the pen. Suppose, for instance, Fig. 48 is pencil sketched, and we are to ink it. First take the sill lines, then the top ridge line, and after that the outside rafter and post lines. This will give the outline of the entire figure. As the rafters are joined by means of a ridge board and rest on top of the purlin and main plates, while at the same time extending below the main plate, it is necessary to next draw the rafters. First draw the ends, then the length lines, after which the roof can be finished by drawing the lower lines of the ridge board and the plate lines as they would appear between the rafters, thus showing just where to start and stop the short lines. The short lines running perpendicularly from the main plate to the purlin plate show the portion of the purlin posts exposed to view at the sides of the rafters on account of the posts being thicker. Next draw the main post lines, girt lines, door posts, headers and braces in the order named. A little practice in this kind of work is the best experience, and as the work progresses step by step the best ways and means of accomplishing certain results will come to light and appear plainer and plainer as the draftsman gains in the knowledge of his profession.

The elevation of the left end bent is shown in Fig. 49, which is self explanatory. A few figures are included to give an idea of marking sizes of timber used. The elevation of the next bent, which shows some changes in the framing, is represented in Fig. 50. This being an inside bent, it is not necessary that the tie girts, to which it is usual to nail the siding, should be in the frame; in fact, they would only be in the way, so for this reason they are omitted. The purlin girt is left out because it would be in the way of operating a hay fork if one was desired. As this bent is the one directly in front of the horses and next to the barn floor it is necessary that it should be boarded up about 3 feet high. For this purpose tie girts are put in and studded underneath, as shown.

It would now appear that the subject of architectural drawing had been considered

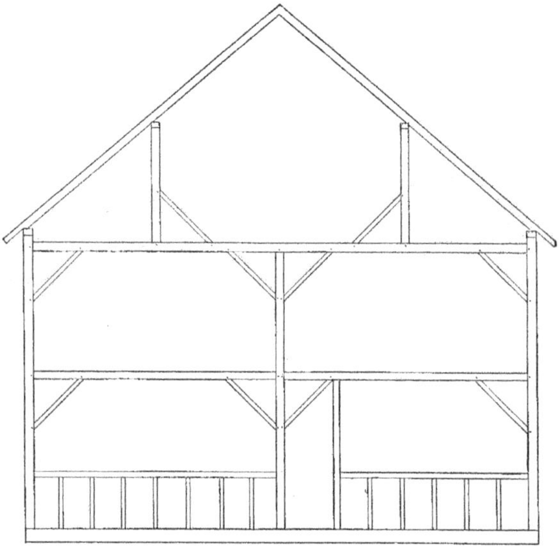

Fig. 50.—Method of Framing and Drawing Middle Bent.—Scale, 1/8 Inch to the Foot.

with sufficient clearness to enable the student to proceed understandingly in ordinary practice and make the working plans so often required in the building trade.

In conclusion, it may be well to say to those who wish to improve themselves in this subject: study the plans, elevations and details of the buildings that appear from time to time in architectural journals. Many of them show specimens of the finest architectural work, which will serve as the very best of lessons for study and practice. Study the work; study how to improve upon it; study the best manner to proceed, and then try a hand at executing the work by making a few drawings for practice.

Very small curves, scrolls and ornaments can be best made with a very fine writing pen, as the drawing pen is not adapted to the short and crooked lines forming the curves of molded surfaces, brackets and small ornaments appearing in elevations.

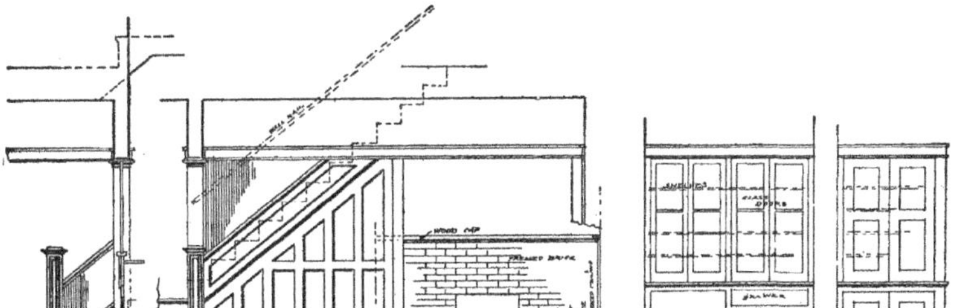

STAIRWAY MANTEL AND SEATS PANTRY CUPBOARD KITCHEN CUPBOARD

Architecture is a study of itself, and the successful architect should have nothing to divert his attention from his work, as it requires undivided attention, and to accomplish the best results he should not work too many hours a day, but be able to work somewhat at his pleasure, say from six to eight hours a day. It is wearisome to the mind to work continuously at it without proper rest. Architectural drawing is a profession which seems to grow in demand and one that still commands a good remuneration for the time consumed in the work. The present rates for architectural services are based as follows: For drawing plans, from 1½ to 2½ per cent. of the cost of construction, and 5 per cent. for personal superintendence. It will be readily seen from this that an architect would receive for drawing plans for a $2,000 residence from $35 to $50, and for plans and personal superintendence $100. Considering that such plans could be prepared in three or four days, and that the personal superintendence is only occasional visits of a few hours each in looking over and inspecting the work as it progresses, it is a pretty fair consideration for services rendered. Yet it is not a higher price than a skillful architect and superintendent should have for such services. There is plenty of encouragement for young architects. The work is pleasant, light and profitable. Remember that there is always room at the top, that the demand for skilled workmen is constantly increasing and that it is the class of unskilled workmen who are the most unemployed. This should be an incentive to every student, professional man and tradesman to aspire to higher qualifications and attainments in his particular line of business. Skill and talent combined with practice and experience is the one essential qualification which leads to success.

We trust that the instructions we have given in this work will serve the purpose for which they were intended and give a start to those seeking information that will eventually bring them to a thorough understanding and successful issue on the subject set forth.

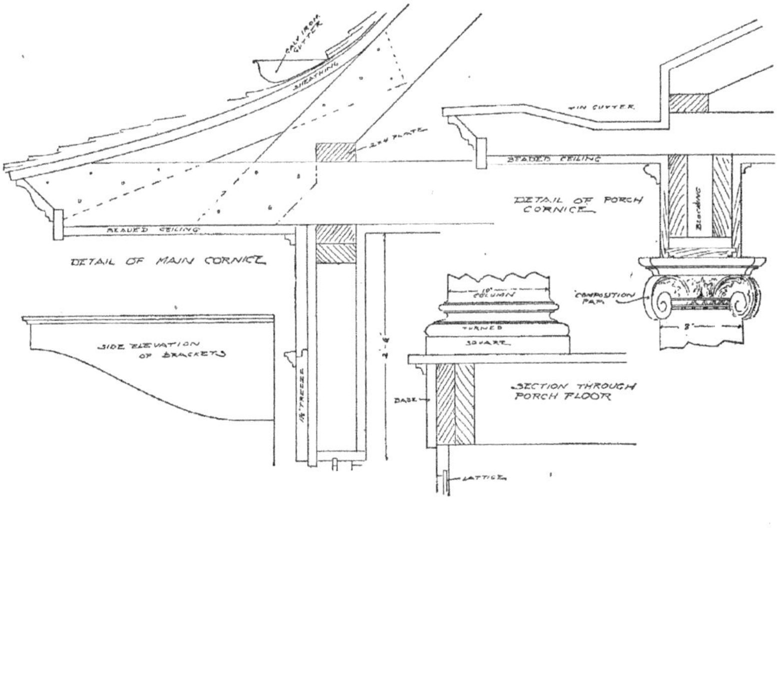
GALV IRON GUTTER
SHEATHING
2x4 PLATE
BEADED CEILING
DETAIL OF MAIN CORNICE
TIN GUTTER
BEADED CEILING
DETAIL OF PORCH CORNICE
Blocking
COMPOSITION CAP
SIDE ELEVATION OF BRACKETS
10" COLUMN
TURNED
SQUARE
SECTION THROUGH PORCH FLOOR
BASE
LATTICE

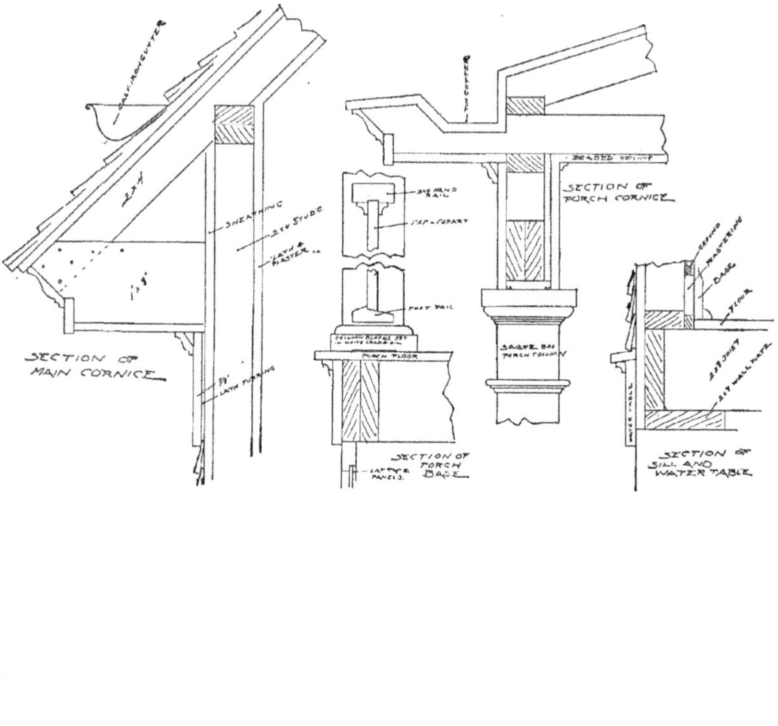
SHEATHING
2x4 STUDS
LATH & PLASTER
2x4
LATH TURNING
SECTION OF
MAIN CORNICE
TIN GUTTER
BEADED CEILING
SECTION OF
PORCH CORNICE
HAND RAIL
FOOT RAIL
PORCH FLOOR
SQUARE BOX
PORCH COLUMN
LATTICE
PANELS.
SECTION OF
PORCH
BASE
GROUND
PLASTERING
BASE
FLOOR
WATER TABLE
SECTION OF
SILL AND
WATER TABLE

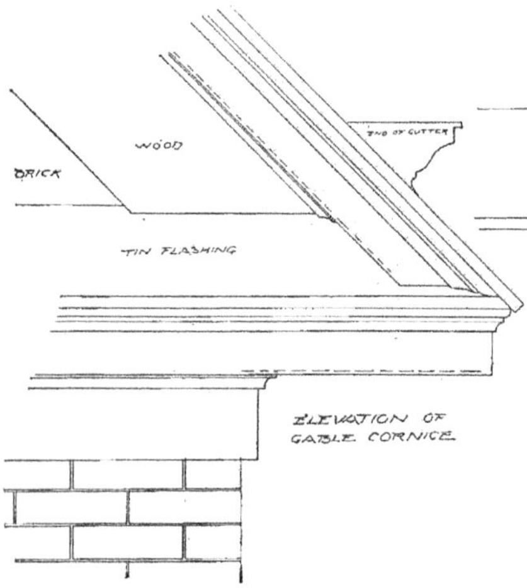

ELEVATION OF GABLE CORNICE

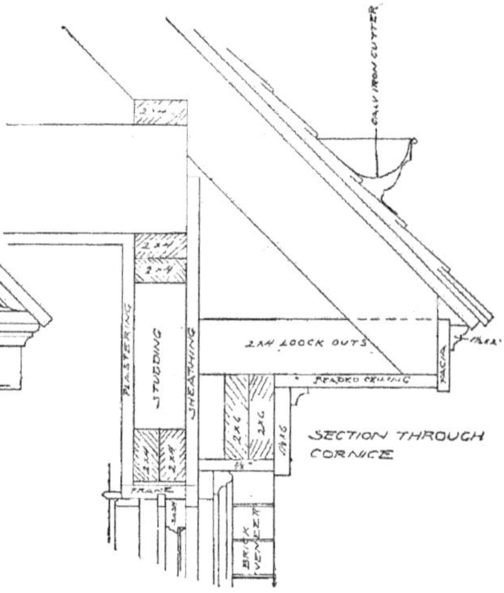

SECTION THROUGH CORNICE

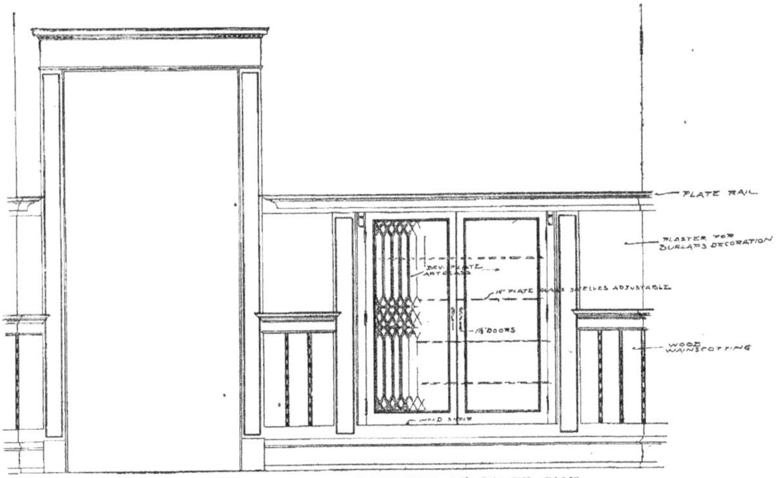

ELEVATION OF SILVER CASE

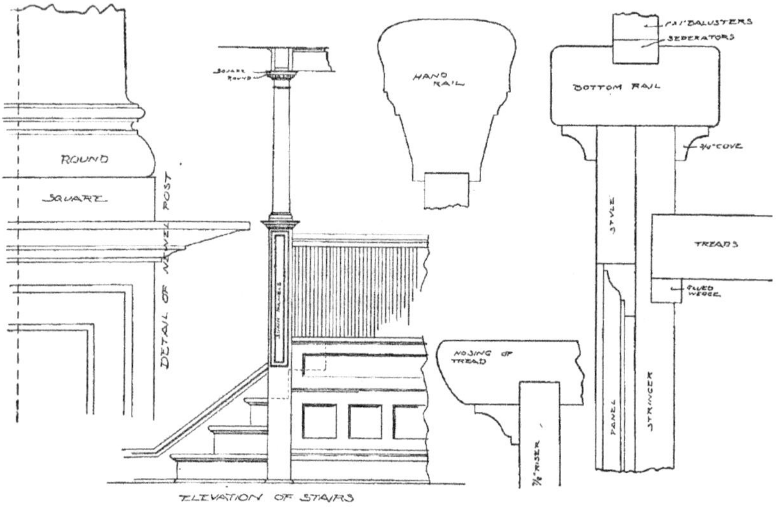
ROUND
SQUARE
DETAIL OF NEWEL POST
HAND RAIL
1 x 1" BALUSTERS
SEPERATORS
BOTTOM RAIL
3/4" COVE
STYLE
TREADS
GLUED WEDGE
NOSING OF TREAD
7/8" RISER
PANEL
STRINGER
ELEVATION OF STAIRS

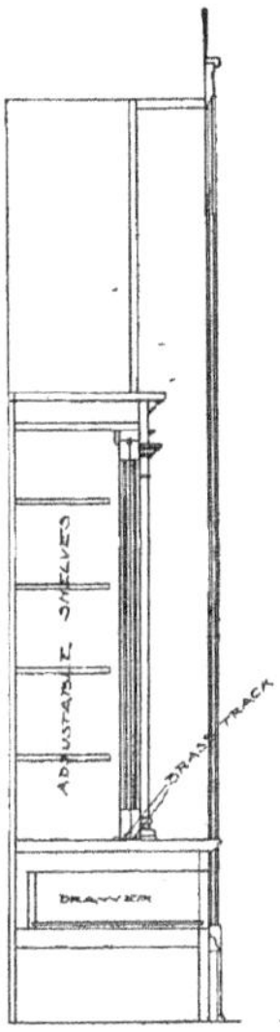

SECTION

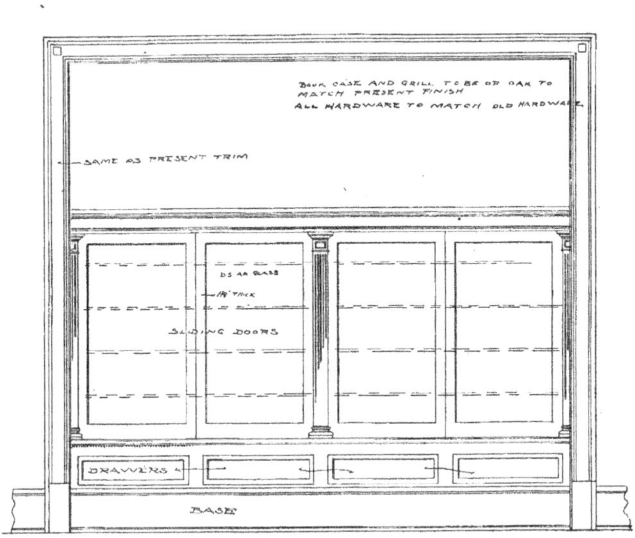

ELEVATION OF BOOK CASE

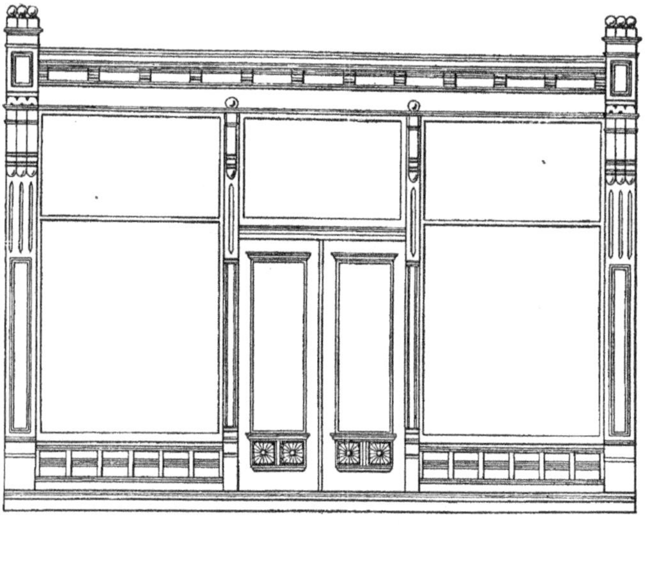

Spotted something that isn't quite right?

Let us know and we'll fix it for future reprints. And as a thank you for your diligence, we'll send you a free Sothis book of your choice.

publisher@sothis.press

This page intentionally left blank

Made in the USA
Monee, IL
08 November 2026

68106063R00052